T0205241

METHODS IN MOLECULAR BIOLOGY

Series Editor
John M. Walker
School of Life and Medical Sciences
University of Hertfordshire
Hatfield, Hertfordshire, AL10 9AB, UK

For further volumes:
http://www.springer.com/series/7651

Antibiotic Resistance Protocols

Third Edition

Edited by

Stephen H. Gillespie

School of Medicine, University of St. Andrews, North Haugh, UK

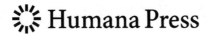 Humana Press

Editor
Stephen H. Gillespie
School of Medicine
University of St. Andrews
North Haugh, UK

ISSN 1064-3745   ISSN 1940-6029   (electronic)
Methods in Molecular Biology
ISBN 978-1-4939-8534-0   ISBN 978-1-4939-7638-6   (eBook)
https://doi.org/10.1007/978-1-4939-7638-6

Printed on acid-free paper

This Humana Press imprint is published by Springer Nature
The registered company is Springer Science+Business Media, LLC
The registered company address is: 233 Spring Street, New York, NY 10013, U.S.A.

Preface

Since the first two editions in this series were published, much has changed in the world of antibiotic resistance research. Importantly, national and international reports have highlighted the cost of antibiotics resistance to world health. There is a recognition of the need to understand the processes that drive resistance and to identify ways to ameliorate the apparently inexorable progress to a post-antibiotic age.

At the same time, our science has not stood still and there are new tools to address the old questions. This new edition of *Antibiotic Resistance Protocols* draws on a wide range of different technologies. These range from conventional growth-based techniques to the application of molecular biology to understand the development of resistance mutations, diagnosis, and monitoring treatment response. Importantly, we are now starting to see beyond genetic resistance to start to understand how transitory phenotypic resistance may play a role in the emergence of fully resistant strains. Within this volume there are techniques from the microscopic scale to whole animal models.

Antibiotic resistance is, truly, a major threat to modern medicine, which is only possible with the contribution of antibiotics to support patients through complex procedures. It is only by redoubling our research efforts are we likely to address this problem, and the third edition of this series provides tools that we hope you will find useful in your contribution to progress.

St. Andrews, UK   *Stephen H. Gillespie*

Contents

Contributors

NIELS AGERSNAP • *Philips BioCell A/S, Allerød, Denmark*

MANUEL ALCALDE-RICO • *Departamento de Biotecnología Microbiana, Centro Nacional de Biotecnología, CSIC, Madrid, Spain*

SABA ANWAR • *TB Research, National Infection Service, Public Health England, Salisbury, UK*

JOANNA BACON • *Research, National Infectious Service, Public Health England, Salisbury, UK*

VINCENT O. BARON • *School of Medicine, University of St Andrews, St Andrews, UK*

CHRISTOPHER BURTON • *TB Research, National Infection Service, Public Health England, Salisbury, UK*

CHIARA CANALI • *Philips BioCell A/S, Allerød, Denmark*

MINGZHOU CHEN • *SUPA, School of Physics and Astronomy, University of St Andrews, St Andrews, UK*

SIMON O. CLARK • *National Infectious Service, Public Health England, Salisbury, Wiltshire, UK*

ANTHONY COATES • *Institute of Infection and Immunity, St George's, University of London, London, UK*

KISHAN DHOLAKIA • *SUPA, School of Physics and Astronomy, University of St Andrews, St Andrews, UK*

MARINA ELEZ • *iSSB, Genopole, CNRS, UEVE, Université Paris-Saclay, Évry, France; LJP, CNRS UMR 8237, UPMC, Sorbonne Universités, Paris, France*

SIMON J. FOSTER • *Krebs Institute, University of Sheffield, Sheffield, UK; Department of Molecular Biology and Biotechnology, University of Sheffield, Sheffield, UK; Bateson Centre, University of Sheffield, Sheffield, UK*

NIELS FRIMODT-MØLLER • *Department of Clinical Microbiology, Rigshospitalet, Copenhagen, Denmark*

STEPHEN H. GILLESPIE • *School of Medicine, University of St Andrews, St Andrews, UK*

KATHERINE A. GOULD • *Institute for Infection and Immunity, St George's University of London, London, UK*

ROBERT J.H. HAMMOND • *School of Medicine, University of St Andrews, St Andrews, UK*

CHARLOTTE L. HENDON-DUNN • *TB Research, National Infection Service, Public Health England, Salisbury, UK*

FREDERIK BOËTIUS HERTZ • *Hvidovre Hospital, Hvidovre, Denmark; Statens Serum Institut, Copenhagen, Denmark*

YANMIN HU • *Institute of Infection and Immunity, St George's, University of London, London, UK*

DIARMAID HUGHES • *Department of Medical Biochemistry and Microbiology, Uppsala University, Uppsala, Sweden*

DOUGLAS L. HUSEBY • *Department of Medical Biochemistry and Microbiology, Uppsala University, Uppsala, Sweden*

Sam Lipworth • *Department of Tropical Medicine and Global Health, University of Oxford, Oxford, UK*

José Luis Martínez • *Departamento de Biotecnología Microbiana, Centro Nacional de Biotecnología, CSIC, Madrid, Spain*

Ivan Matic • *INSERM U1001, Université Paris-Descartes, Sorbonne Paris Cité, Faculté de Médecine Paris Descartes, Paris, France; Centre National de la Recherche Scientifique (CNRS), Paris, France*

Gareth McVicker • *Krebs Institute, University of Sheffield, Sheffield, UK; Department of Molecular Biology and Biotechnology, University of Sheffield, Sheffield, UK; Bateson Centre, University of Sheffield, Sheffield, UK*

Karen Leth Nielsen • *Statens Serum Institut, Copenhagen, Denmark; Department of Clinical Microbiology, Rigshospitalet, Copenhagen, Denmark*

Tom Olesen • *Philips BioCell A/S, Allerød, Denmark; BioSense Solutions ApS, Farum, Denmark*

Tomasz K. Prajsnar • *Krebs Institute, University of Sheffield, Sheffield, UK; Department of Molecular Biology and Biotechnology, University of Sheffield, Sheffield, UK; Bateson Centre, University of Sheffield, Sheffield, UK*

Stephen A. Renshaw • *Krebs Institute, University of Sheffield, Sheffield, UK; Bateson Centre, University of Sheffield, Sheffield, UK; Department of Infection, Immunity and Cardiovascular Disease, University of Sheffield, Sheffield, UK*

Lydia Robert • *LJP, CNRS UMR 8237, UPMC, Sorbonne Universités, Paris, France; Micalis Institute, INRA, AgroParisTech, Université Paris-Saclay, Jouy-en-Josas, France*

Erik Spillum • *Philips BioCell A/S, Allerød, Denmark; BioSense Solutions ApS, Farum, Denmark*

Stephen C. Taylor • *Research, National Infectious Service, Public Health England, Salisbury, UK*

Stephen R. Thomas • *Research, National Infectious Service, Public Health England, Salisbury, UK*

Mark A. Toleman • *Department of Infection and Immunity, Cardiff University, Cardiff, UK*

Martin Valvik • *Philips BioCell A/S, Allerød, Denmark*

Simon J. Waddell • *Department of Global Health and Infection, Brighton and Sussex Medical School, University of Sussex, Brighton, UK*

Leticia Muraro Wildner • *Universidade Federal de Santa Catarina, Florianópolis, Santa Catarina, Brazil; Department of Global Health and Infection, Brighton and Sussex Medical School, University of Sussex, Brighton, UK*

Alexander Williams • *Krebs Institute, University of Sheffield, Sheffield, UK; Department of Molecular Biology and Biotechnology, University of Sheffield, Sheffield, UK; Bateson Centre, University of Sheffield, Sheffield, UK*

Ann Williams • *National Infectious Service, Public Health England, Salisbury, Wiltshire, UK*

Han Xaio • *School of Medicine, University of St Andrews, St Andrews, UK*

Chapter 1

Methods for Measuring the Production of Quorum Sensing Signal Molecules

Manuel Alcalde-Rico and José Luis Martínez

Abstract

One relevant aspect for understanding the bottlenecks that modulate the spread of resistance among bacterial pathogens consists in the effect that the acquisition of resistance may have on the microbial physiology. Whereas studies on the effect of acquiring resistance of bacterial growth are frequently performed, more detailed analyses aiming to understand in depth the cross talk between resistance and virulence, including bacterial communication are less frequent. The bacterial quorum sensing system, is an important intraspecific and interspecific communication system highly relevant for many physiological processes, including virulence and bacterial/host interactions. Some works have shown that the acquisition of antibiotic resistance may impair the quorum sensing response. In addition, some antibiotics as antimicrobial peptides can affect the production and accumulation of the quorum sensing signal molecules. Given the relevance that this system has in the bacterial behavior in the human host, it is important to study the effect that the acquisition of antibiotic resistance may have on the production of quorum sensing signals. In this chapter we present a set of methods for measuring quorum sensing signals based on the use of biosensor strains, either coupled to Thin Layer Chromatography or for performing automated luminometry/spectrophotometry assays. We use *Pseudomonas aeruginosa* as bacterial model because it has a complex quorum system than encloses different signals. Namely, *P. aeruginosa* quorum sensing system consists in three different interconnected regulatory networks, each one presenting a specific autoinducer molecule: the *las* system, which signal is *N*-(3-oxo-dodecanoyl)-L-homoserine lactone, the *rhl* system, which signal is *N*-butanoyl-homoserine lactone and the *pqs* system, which signals are 2-heptyl-3-hydroxy-4(1H)-quinolone together with its immediate precursor 2-heptyl-4-hydroxy-quinoline.

Key words Quorum sensing, *Pseudomonas aeruginosa*, Pseudomonas Quinolone Signal, 2-Alkyl-4(1H)-quinolones, *N*-acyl homoserine lactones, Antibiotic resistance, Thin layer chromatography, PQS, AHL, Fitness cost

1   Introduction

It is generally assumed that the acquisition of resistance is associated with fitness costs that make resistant bacteria to be less proficient for growing in different ecosystems than their wild-type counterparts. Whilst this is true in occasions [1–3], in other cases, the acquisition of resistance produces specific changes in the bacterial physiology that do not correlate with growth alter-

Stephen H. Gillespie (ed.), *Antibiotic Resistance Protocols*, Methods in Molecular Biology, vol. 1736,
https://doi.org/10.1007/978-1-4939-7638-6_1, © Springer Science+Business Media, LLC 2018

ations [4]. Among such changes, one of relevance for bacterial behavior consists on alterations in the quorum sensing (QS) response. Indeed, different articles have shown that some multidrug efflux pumps are able to extrude QS signal molecules (QSSMs) or their metabolic precursors. This situation makes that the acquisition of resistance due to due to the overexpression of these efflux pumps might impair the QS response, and consequently the expression of virulence factors, of resistant strains [5–10]. The QS system is dependent on cell density; serves to determine the cells concentration in a given environment. When the population density reaches a given threshold, a coordinated response is triggered, which is relevant for several bacterial processes including virulence [11, 12]. The process starts with the production of one or more low molecular weight compounds, known as "autoinducers," by the cells and their diffusion across the membrane. This diffusion results in a progressive accumulation of the signal until the signal threshold level needed to produce the signaling cascade is reached [13, 14]. This quick response is due to the efficient binding of the QSSM to its specific transcriptional regulator, which activates (or represses) the transcription of a high number QS-regulated genes [11, 12, 15, 16]. In general, among those genes positively regulated by QS are both the transcription factors that mediate the QS response and the enzymes that catalyze the synthesis of the QSSMs resulting in a positive feedback regulation [16, 17]. For this reason the QSSMs are named autoinducers too.

Since the first QS phenomenon was discovered by Nealson in 1970 [18, 19], a large number of signaling properties have been associated to several small molecules synthesized by different bacteria [17, 20–22]. The best studied signals in Gram-negative bacteria are the *N*-acyl homoserine lactones (AHLs) and the 2-alkyl-4(1H)-quinolones (AQs), whereas in Gram-positive bacteria the most important ones are the autoinducer peptides (AIPs). The different QS networks regulated by these autoinducers drive the expression of several genes involved in many biological processes as the production of antibiotics and virulence factors (elastase, proteases, siderophores, toxins, phenazines, T3SS or T6SS), biofilm maturation, bioluminescence, swarming motility, sporulation or antibiotics resistance, among others [11, 12, 14]. In addition to their role in mediating the communication among members of the same species, QS also may be involved in interspecific and even interkingdom signaling [16, 17, 23, 24].

The common structure of AHLs autoinducers is a homoserine lactone ring attached to an acyl chain through an amide bond. The number of carbon atoms of this acyl chain can vary and the third position may be modified in occasions with carbonyl or hydroxyl group [25–27]. These different structures made AHLs sufficiently different to be recognized by specific sensor proteins of the LuxR family [27]. One of the best characterized AHLs signaling system is based on N-(3-oxo-dodecanoyl)-L-homoserine lactone (3-oxo-C12-HSL), which is involved, in *P. aeruginosa*, in both intraspecific signaling (i.e., elastase and exoproteases production) and interspecific signaling (i.e., with yeast or mammalian cells) [11, 23, 24]. One relevant interkingdom signaling process of this autoinducer is the interaction with human cells. Smith et al. [28] have shown that 3-oxo-C12-HSL induces the production of several pro-inflammatory chemokines, including IL-8 in human bronchiolar epithelial cells and lung fibroblast. In addition, 3-oxo-C12-HSL can act as chemoattractant for neutrophils inducing their migration to the site where the signal is released [29, 30].

In addition, to their regulatory role, QSSMs may have nonsignaling properties, including antibiotic activity or iron chelation. A good example of this antibiotic activity is shown by the lantibiotics (i.e., nisin produced by *Lactobacillus lactis* or subtilin produced by *Bacillus subtilis*), a group of antimicrobial compounds that are closely related with AIPs [16]. In the same line of reasoning, the 2-heptyl-3-hydroxy-4(1H)-quinolone (*Pseudomonas quinolone signal*, PQS) is used by *P. aeruginosa* to capture iron when growing inside an infected host, and also to steal the iron stores of other bacteria [31].

All these effects on bacterial physiology and virulence support the need of establishing clear and robust protocols to detect the QSSMs, and determine the effect of antibiotic resistance on the production of such compounds. The most useful techniques for such purpose are based on the use of biosensor strains. These biosensors do not produce any QSSMs but contain the sensor protein that recognizes the autoinducer of interest. The complex formed by QSSM and the sensor protein promotes the transcription of a reporter gene, producing a detectable signal, including bioluminescence, fluorescence, pigments production or β-galactosidase activity [32]. It is important considering the limitations of this technique, specially when working with bacterial species in which the studies on QS and on the corresponding QSSMs are scarce or even absent. In particular, it is important: (1) to know the minimal and saturating concentration of autoinducer to produce the signal of each biosensor strain, (2) to carry out positive and negative controls in the experiment to address the possibility of the presence of either quenchers or enhancers of the system under the studied conditions, (3) to establish specific conditions in which production of the QSSM to be studied is granted. In this protocol, we use as models for their detection the QSSMs produced by *P. aeruginosa* [11, 15]. The method is based on the use of Thin Layer Chromatography combined with a biosensor overlay [25, 26, 33–35]. In addition, we also describe how to quantify QSSMs with the use of an automated luminometer-spectrophotometer [26, 27, 35, 36]. More details concerning other AHLs bacterial biosensors, not described here, can be found in [32].

2 Materials

2.1 Bacterial Strains

1. *Pseudomonas aeruginosa* PAO1 wild type.

2. *P. aeruginosa* Δ*pqsA*.

3. *Escherichia coli* pSB1705 (LasR-based bioreporter) [27].

4. *E. coli* pSB536 (RhlR-based bioreporter) [37].

5. *P. aeruginosa* PAO1 *pqsA* CTX-*lux::pqsA* (PqsR-based bioreporter) [33].

2.2 Growth Media

1. Luria-Broth (LB).

2. LB agar plates: LB medium with bacteriological agar at 1.5% ($^w/_v$). Pour in each petri plate 20 ml of this LB agar.

3. LB agar to overlay: LB medium with bacteriological agar at 0.75% ($^w/_v$).

4. Soft top agar medium 0.65% ($^w/_v$): dissolve in H_2O milliQ agar 0.65% ($^w/_v$), tryptone 1% ($^w/_v$), and sodium chloride (NaCl) 0.5% ($^w/_v$).

2.3 Reagents

1. Acidified ethyl acetate HPLC grade 0.01% ($^v/_v$): add glacial acetic acid to ethyl acetate HPLC grade at a final concentration 0.01% ($^v/_v$).

2. Methanol HPLC grade.

3. Glacial acetic acid.

4. Dichloromethane HPLC grade.

5. Acetone for analysis.

6. Potassium dihydrogen phosphate solution 5% ($^w/_v$): KH_2PO_4 dissolved in H_2O_d.

7. Synthetic QS signal molecules: AQs suspend in methanol HPLC grade and AHLs suspend in ethyl acetate HPLC grade.

8. Ampicillin stock 100 mg/ml dissolved in H_2O milliQ.

9. Tetracycline stock 10 mg/ml dissolved in ethanol 70% for analysis.

10. H_2O milliQ (18.2 MΩ·cm at 25 °C).

2.4 Equipment

In most cases, general equipment can be purchased from different companies and a specific reference to the model used in our laboratory is not included.

1. Thin layer chromatography silica gel 60 F_{254} 20 cm × 20 cm (AQs).

2. Thin layer chromatography silica gel 60 RP-18 F_{254s} 20 cm × 20 cm (C4-HSL).

3. Thin layer chromatography silica gel 60 RP-2 F_{254s} 20 cm × 20 cm (3-oxo-C12-HSL).

4. 50 ml, 100 ml and 250 ml glass flasks.

5. HPLC glass tubes with cover.

6. Centrifuge microtubes 1.5 ml.

7. Centrifuge tubes 50 ml.

8. Spectrophotometer cuvettes.

9. 96-well white flat transparent microtiter plates.

10. Vortex.

11. Microwave oven.

12. Oven.

13. Centrifuge.

14. Shaking incubator.

15. −20 °C freezer, −80 °C freezer, and fridge.

16. Spectrophotometer.

17. Film developer machine.

18. Combined automated luminometer-spectrophotometer.

19. Bunsen burner.

20. Sterile 0.2 μm size filters.

21. 1 ml and 10 ml syringe.

22. High-sensitivity X-ray film as Curix RP2 Plus (Agfa) or BioMax (Kodak).

23. TLC developing tank.

3 Methods

3.1 Analysis of Quorum Sensing Signal Molecules by Thin Layer Chromatography and Biosensor-Based Detection

3.1.1 Culture Conditions to Extract AHLs and AQs Signal Molecules from P. aeruginosa

1. Inoculate 10 ml of LB in a 50 ml flask with a single colony of each strain to be tested from freshly grown agar plates (*see* **Notes 1** and **2**). Grow overnight at 37 °C with shaking at 250 rpm.

2. Next day, determine the optical density at 600 nm (OD_{600}) and dilute the cultures to OD_{600} = 0.01 in 100 ml flasks containing 25 ml of fresh LB medium. Incubate at 37 °C with shaking at 250 rpm.

3. To synchronize the cultures, grow to exponential phase (OD_{600} = 0.5–0.6) and dilute again in two different flasks at OD_{600} = 0.01 to have biological duplicates. Grow the cultures (*see* **Note 3**) until early stationary phase (approximately OD_{600} = 2.0) (*see* **Note 4**).

4. When two replicas of each strain are at the desired OD_{600}, mix them in a flask by gentle agitation and pass 11 ml of each culture to a centrifuge tube. Centrifuge at $6000 \times g$, 4 °C for 15 min.

5. Recover the supernatant carefully to avoid losing cellular pellet and filter it through a sterile 0.2 μm size filter. The obtained

cellular pellet will be used for the extraction of cell-associated autoinducers.

3.1.2 AHLs and AQs Extraction in Cell-Free Supernatant

1. Add 10 ml of the cell-free supernatant to a clean centrifuge tube containing the same volume of acidified ethyl acetate (10 ml) and vortex approximately 30 s until the two phases are completely mixed.

2. Centrifuge at 6000 × *g*, 4 °C for 20 min and transfer 8 ml of the upper organic layer to 250 ml flasks (*see* **Notes 5–7**).

3. Evaporate them each cell-free supernatant with a stream of nitrogen. Alternatively you can also rotary evaporate the extracts.

4. Add 4 ml of acidified ethyl acetate (AHLs extraction) or HPLC grade methanol (AQs extraction) to the flasks and agitate well for a brief time (30 s) to suspend the dry extracts (*see* **Note 7**).

5. Transfer each entire solution in aliquots of 300 µl to glass cap vials (HPLC vials for instance) (*see* **Note 7**), dry under a stream of nitrogen and store them until further use. The Subheading 3.1.2, **steps 4** and **5** can be repeated if you want to get a more exhaustive extraction (*see* **Notes 8** and **9**).

3.1.3 AQs Extraction from the Cellular Fraction (Continue from Subheading 3.1.1, Step 5)

1. Wash the pellet by adding 10 ml of LB fresh medium with a pipette and resuspend it carefully (*see* **Note 10**). Centrifuge at 6000 × *g*, 4 °C for 15 min and discard the supernatant. Repeat this step once.

2. Add 10 ml of HPLC grade methanol and resuspend the pellet by vigorous vortexing (*see* **Note 11**). Wait for 10 min until the methanol lyses the cells completely.

3. Centrifuge at 6000 × *g*, 4 °C for 20 min and transfer the supernatant (methanol containing the AQs molecules) to 250 ml flasks (*see* **Notes 6** and **7**).

4. Evaporate the extracts of each cellular pellet under a stream of nitrogen. Alternatively you can also rotary evaporate the extracts.

5. Add 4 ml of HPLC grade methanol to the flasks and agitate well (30 s) to suspend all extract (*see* **Note 7**).

6. Transfer each entire solution in aliquots of 300 µl to glass cap vials (HPLC vials for instance) (*see* **Note 7**), dry them under a stream of nitrogen and store them until further use. The Subheading 3.1.3, **steps 5** and **6** could be repeated in order to optimize the extraction (*see* **Notes 8** and **9**).

3.1.4 Preparing Reporter Strains Cultures for Overlaying TLC Plates

1. Inoculate 10 ml of LB containing either ampicillin 100 µg/ml (RhlR-based biosensor[pSB536]) [37], tetracycline 5 µg/ml (LasR-based biosensor [pSB1705]) [27] or tetracycline 125 µg/

ml (PqsR-based biosensor) [33] in 50 ml flasks with a single colony of each reporter strain from freshly grown agar plates with their respective antibiotics (*see* **Note 12**). Grow overnight at 30 °C or 37 °C under shaking at 250 rpm (*see* **Note 3**).

2. Next day, melt 100–120 ml of LB agar 0.75% (wt/vol) (AHLs reporter) or soft top agar 0.65% (wt/vol) (PqsR-based biosensor) and temperate at 50 °C.

3.1.5 TLC-Assay
of the Extract
of Supernatants
for Subsequent Detection
of C4-HSL or
3-Oxo-C12-HSL

1. Draw carefully a line with a pencil at 2 cm from the edge of the TLC plate (RP-18 F254s for C4-HSL and RP-2 F254s for 3-oxo-C12-HSL) and mark the points where the samples will be loaded (*see* **Note 13**).

2. Suspend the sample extracts of Subheading 3.1.2, **step 5** in 1 ml of acidified ethyl acetate and spot 5 μl of each on its place on the TLC plate (*see* **Note 14**), trying the spots to be as smallest as possible by drying samples during spotting (*see* **Note 15**).

3. Synthetic C4-HSL or 3-Oxo-C12-HSL dissolved in acidified ethyl acetate should be used as positive controls. Spot 1–2 μl of 10 μM of this solution in the TLC plate. The minimal distance recommended for optimal resolution is 2 cm between two spots (*see* **Note 14**).

4. When spots are dried, put the TLC plate into a developing chamber that contains 150 ml of the mobile phase mixture, methanol–water (60:40 [vol/vol] for C4-HSL and 45:55 [vol/vol] for 3-Oxo-C12-HSL) (*see* **Note 16**). Run the TLC by capillarity until the solvent reaches 2–4 cm from the top of the plate and then let it dry completely inside the fume hood (*see* **Note 17**).

3.1.6 TLC-Assay
of the Extract of Both
Cell-Fraction
and Supernatants
for Subsequent Detection
of PQS
and 2-Heptyl-4-Hydroxy-
Quinolone HHQ

1. Activate the TLC sheets (silica gel 60 F_{254}): prepare 1 l of KH_2PO_4 solution 5% ($^{wt}/_{vol}$) in distilled water and transfer to a wide clean tray. Immerse the TLC sheets in KH_2PO_4 solution for 30 min at room temperature (*see* **Note 18**). Then, introduce the TLC plates into a prewarmed oven avoiding them to contact one to each other and keep them at 100 °C for 1 h (*see* **Note 19**). Once you have activated the TLC plate, you should label the position of the spots as explained above in Subheading 3.1.5, **step 1**.

2. Suspend the sample extracts of Subheading 3.1.2, **steps 5** and **6** in 100 μl of methanol HPLC grade and spot 30 μl of each on its place in the TLC sheet (*see* **Notes 14** and **15**).

3. Synthetic PQS and HHQ dissolved in methanol HPLC should be used as positive control. Spot 1–2 μl of 10 mM of each solution and a mixture of both in TLC plate (*see* **Notes 14** and **20**).

4. When spots are dried, put the TLC plate into a developing chamber that contains 100 ml of the mobile phase; dichloromethane–methanol (95:5 [vol/vol]) (*see* **Note 16**). Run the

TLC until the solvent reaches 2–4 cm from the top of the plate and let dry (*see* **Note 17**).

3.1.7 Biosensor-Based Detection of AHLs or AQs Signal Molecules in TLC-Assay

1. Once the TLC plate obtained after Subheadings 3.1.5, **step 4** or 3.1.6, **step 4** is dried, make a pool with a depth of at least 0.5 cm by rounding the TLC plate with autoclaved tape (*see* **Note 21**).

2. Inoculate the agar-containing medium from Subheading 3.1.4, **step 2** with 1.0–1.2 ml of the overnight biosensor culture obtained as described in Subheading 3.1.4, **step 1** (*see* **Note 22**) and mix carefully to avoid bubble formation (*see* **Note 23**).

3. Pour the inoculated medium over the preformed pool slowly taking care that the entire TLC plate is covered homogeneously (*see* **Note 24**). Wait until the agar is completely solidified under sterile conditions.

4. Incubate the plates 14–16 h at 30 °C for the AHLs-based bioassay or 6 h at 37 °C for the PqsR-based bioassay.

5. After incubation, you can use a luminograph photon video camera (for instance Luminograph LB 980 photon video camera from Berthold Technologies USA) to visualize the light emitted in the bioassay.

6. Alternatively, you can expose an X-ray film over the plate: put the plate in an X-ray chamber, remove with a scalpel or a razor blade the border of the tape carefully and cover the agar plate with a plastic transparent film. Place the X-ray film over the agar plate in darkness, close the X-ray chamber and expose about 2–10 min depending on the identity of the signal (*see* **Note 25**).

7. To develop the X-ray film, use a regular developer machine (Fig. 1).

3.2 Analysis of the Kinetics of QS Signal Molecules Accumulation into Supernatant Along the Cell Cycle by a Method Based in a Combined Automated Luminometer-Spectrophotometer

1. Inoculate 10 ml of LB containing either ampicillin 100 μg/ml (RhlR-based biosensor[pSB536]), tetracycline 5 μg/ml (LasR-based biosensor [pSB1705]) or tetracycline 125 μg/ml (PqsR-based biosensor) in 50 ml flasks with a single colony of each reporter strain from freshly grown agar plates with their respective antibiotics (*see* **Note 12**). Grow overnight at 30 °C or 37 °C under shaking at 250 rpm (*see* **Note 3**).

2. The next day, dilute 1:100 the overnight biosensor culture in fresh LB medium and incubate the biosensor cells with QSSMs-containing extracts in a 96-well plate as described below (Subheading 3.2.2, **step 9**).

3.2.1 Preparing Reporter Strains Cultures for AHLs and AQs Detection by 96-Well Plate Bioassay

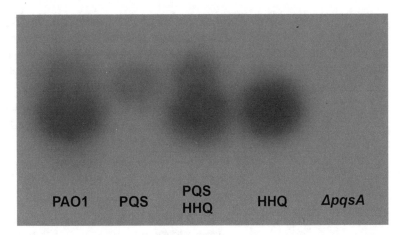

Fig. 1 TLC image of PQS and HHQ extracted from *P. aeruginosa* supernatant culture. Positive controls were included as described in methods: 2 µl from PQS-methanol 10 mM, 2 µl from HHQ-methanol 10 mM, and 2 µl from the combination of both (1 µl from each one 10 mM PQS and 10 mM HHQ). The extraction of AQs from PAO1 and *ΔpqsA* strains was performed at 8 h of incubation at 37 °C with shaking (250 rpm) and 20 µl from each extract was spotted on the TLC plate. The mobile phase used was dichloromethane–methanol (95:5). The time of incubation with de biosensor PAO1 *pqsA* CTX-*lux::pqsA* was 6 h

3.2.2   AHLs and AHQs Extraction in Cell-Free Supernatant from P. aeruginosa Culture and Detection for 96-Well Plate Bioassay

1. Inoculate 10 ml of LB in a 50 ml flask with a single colony of each strain to be tested from freshly grown agar plates (*see* **Notes 1** and **2**). Grow overnight at 37 °C with shaking at 250 rpm.

2. Next day, determine the OD_{600} and dilute the cultures to exponential phase ($OD_{600} = 0.01$) in 100 ml flasks containing 25 ml of fresh LB medium. Incubate at 37 °C with shaking at 250 rpm.

3. To synchronize the cultures, grow to $OD_{600} = 0.5$–0.6 and dilute again at $OD_{600} = 0.01$. Incubate at 37 °C with shaking at 250 rpm. The QS signal molecules will be extracted at different times after inoculation from the same culture. The number of extractions from each culture should be at least four for getting a robust graphic representation of results (*see* **Note 26**).

4. Transfer 1.5 ml of each culture into a microcentrifuge tube and centrifuge at $10,000 \times g$, 4 °C for 5 min. At the same time, determinate the OD_{600} of each culture. Recover the supernatants and filter them through a sterile 0.2 µm size filter.

5. To extract the AHLs signal molecules, add 900 µl of these cell-free supernatants to previously labeled microcentrifuge tubes each one containing 100 µl of HCl 1 M and mix well with a micropipette. Store the rest of the filtered supernatant (600 µl) in the freezer at −20 °C for the subsequent AQs analysis.

6. When all AHLs extractions have been performed incubate them at 20 °C for 16–18 h (overnight) with shaking at 250 rpm (*see* **Note 27**).

7. Next day, 5 µl from each sample (AHLs or AQs extract) is charged into each well of a microtiter plate. Include in the same plate at least three biological replicates of each strain and two technical replicates of each extract (*see* **Note 28**).

8. You can make a standard curve for C4-HSL and 3-oxo-C12-HSL adding 5 µl from 6 or more different concentration solutions of these autoinducers suspended with acidified ethyl acetate (*see* **Note 29**).

9. Add to each test well 195 µl of biosensor culture diluted from Subheading 3.2.1, **step 2**.

10. Include into the bioassay control wells with the bioreporter strain without quorum sensing signal molecules to measure autoluminescence and fresh LB medium without bacterium neither quorum sensing signal molecules as blank.

11. Incubate in a rocking platform at 30 °C or 37 °C with shaking at 200 rpm for 6 h (*see* **Note 3**). Every hour, the OD_{600} and bioluminescence are measured using a combined automated luminometer-spectrophotometer.

3.2.3 Analysis of the Results Obtained from Combined Automated Luminometer-Spectrophotometer 96-Well Plate Bioassay

1. Normalize the results by subtracting the background value obtained in the blank wells. Then, obtain the Relative light units (RLU) for both each well and each time by determining the bioluminescence/OD_{600} ratio.

2. Make a graphic representation of RLU variation as a function of to time and choose the optimal time at which RLUs are bigger but and technical replicas do not diverge too much.

3. The values of RLUs obtained at the optimal time point are used to make the graphic representation of RLU variation respect to OD_{600} of each tested strain in which the quorum sensing signal molecules have been extracted from cell-free supernatants (Fig. 2) (*see* **Note 30**).

4. To estimate the concentration of each QS signal molecule in every sample, make a standard curve using the RLU values obtained upon exposure of the biosensor strains to increasing concentrations of the different positive controls.

4   Notes

1. In all cases you have to work with the strains to be tested and with control strains, each one of the latter presenting a known pattern of signals production. In the case of *P. aeruginosa*, the wild type PAO1 strain and mutants in the QS-signaling pathways as *ΔlasI*, *ΔrhlI*, or *ΔpqsA* can be used.

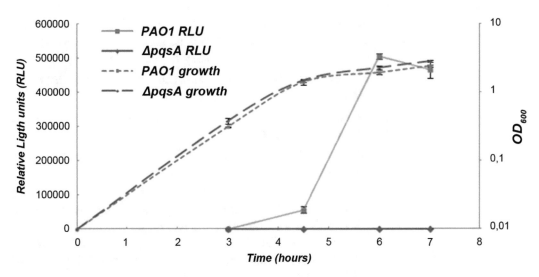

Fig. 2 Analysis of the kinetic AQs accumulation into the supernatants from *P. aeruginosa* PAO1 and *ΔpqsA* strains. The time of incubation from each extraction was 3 h, 4 h 30 min, 6 h and 7 h. The growth curve for each culture is represented in logarithmic scale together with the AHQs accumulation. The RLU values have been calculated for both each well and each time by determining the bioluminescence/OD_{600} ratio

2. To obtain a representative result, you have to repeat the extraction and TLC-bioassay at least with three different biological replicas of each strain. For 96-well plate bioassay you also need 2–3 technical replicas of each sample extraction. However, you can analyze all biological and technical replicas in the same microtiter plate.

3. The temperature used to grow AHLs reporters is 30 °C and for AQs reporter, control strain of experiment and test bacterium is 37 °C.

4. In this example, the growth media used for the extractions of QS signal is LB. However, you can use another media if needed. In the same way, the extraction process can be performed at any stage of the cell cycle.

5. For a better extraction, you can add again 8 ml of acidified ethyl acetate to the same centrifuge tube with the sample and repeat the extraction steps.

6. Extracts of supernatant and cell fraction suspended in acidified ethyl acetate and methanol respectively can be transferred to clean centrifuge tubes and frozen at −20 °C and stored for several days until the drying step.

7. All glass recipients used for the extraction have to be previously cleaned with analysis grade acetone and dried well before their use.

8. Be careful with the drying steps with N_2 to avoid splashing the walls and losing sample. If this happens, try to resuspend the extract from the walls and redry the entire sample again.

9. Dry extracts of supernatant and cell fraction can be frozen at -20 °C and stored for several weeks.

10. To resuspend the cellular pellet do not use vortex because the cells can be broken.

11. Make sure that the cellular pellet is homogenously resuspended without any remaining cell clump.

12. The concentrations of antibiotics used for growing the reporter strains are: ampicillin 100 μg/ml for the RhlR-based biosensor (*E. coli* pSB536), tetracycline 5 μg/ml for the LasR-based biosensor (*E. coli* pSB1705 or 1142) and tetracycline 125 μg/ml for the PqsR-based biosensor (*P. aeruginosa* PAO1 *pqsA* CTX-*lux::pqsA*).

13. Remember to label each loading point before spotting to avoid mistakes along the process and be careful with the pencil to avoid breaking the TLC silica plate.

14. You can modify the volume to resuspend the samples, the quantity that will be spotted and the distance recommended between each other according to its concentration.

15. For a good spotting of the extracts you have to use a tip pipette as thin as possible (i.e., you can use a sterile plastic tip pipette normally used to load thin acrylamide gels ad those used in footprinting assays) and drying the samples along the spotting.

16. The volume of the mobile phase mixture used in this example is for a 23 cm × 23 cm × 7 cm. developing chamber. The volume of mobile phase can be modified accordingly to the dimensions of the developing chamber but the proportion of the mixture cannot be changed.

17. During the running step is important that the TLC plate is equilibrated so as the front of the solvent runs as horizontal as possible.

18. You can immerse 4–5 TLC plates at the same time in KH_2PO_4 solution and heat them together in the oven but touching each other the least possible.

19. Silica gel TLC activated sheets can be stored at room temperature for several months.

20. Mix equal volumes of synthetic PQS 10 mM and HHQ 10 mM into a glass tube for obtain a final concentration of 5 mM for each other.

21. The TLC pool has to be completely sealed without any gap between the autoclave tape and the TLC sheet.

22. The overnight cultures of biosensor strains have always to be diluted 1:100 into agar medium overlay for TLC bioassay.

23. Before inoculating the melted agar medium to be used for the overlay, make sure that the temperature of this medium does not exceed 50 °C.

24. Be careful when pouring the overlay medium into the well to avoid bubble formation and if they form, remove them carefully with the flame of a burner. Do not expose the overlay too much time to the flame to avoid killing the cells.

25. To optimize the image, avoid leaving bubbles and creases. Additionally, depending on the signal intensity you can modify the exposure time of X-ray film to obtain the best image possible.

26. In this experiment the times of extraction are 4: T1 = 3 h [exponential phase]; T2 = 4 h [late exponential phase]; T3 = 5 h [early stationary phase]; T4 = 6 h [stationary phase]).

27. When the samples are incubated with HCl the open-ring lactones that have been formed by AHLs hydrolysis are closed. After this incubation, the AHLs samples can be stored in the fridge at 4 °C for weeks.

28. It is recommended to use a single bioreporter strain for microtiter plate assay because the emitted light is different for each of them and using two report strains in the same plate can produce wrong results because of light contamination between them.

29. For AHLs, The different concentration solutions of AHLs should be in a range of 0–20 µM.

30. Common statistic analyses (as T-test for comparing two samples or ANOVA for comparing several samples) are used to determine the statistical significance of the results.

Acknowledgments

The work in our laboratory is supported by grants BIO2011-25255 and BIO2014-54507-R from the Spanish Ministry of Economy and Competitiveness, S2010/BMD2414 (PROMPT) from CAM, Spanish Network for Research on Infectious Diseases (REIPI RD12/0015) from the Instituto de Salud Carlos III, and HEALTH-F3-2011-282004 (EVOTAR) from the European Union. MAR has been the recipient of an FPI fellowship. Special thanks are given to Miguel Cámara, Paul Williams, and Robert Hancock for providing control strains and QSSMs.

References

1. Andersson DI, Levin BR (1999) The biological cost of antibiotic resistance. Curr Opin Microbiol 2:489–493. https://doi.org/10.1016/S1369-5274(99)00005-3

2. Andersson DI, Hughes D (2010) Antibiotic resistance and its cost: is it possible to reverse resistance? Nat Rev Microbiol 8:260–271. https://doi.org/10.1038/nrmicro2319

3. Shcherbakov D, Akbergenov R, Matt T et al (2010) Directed mutagenesis of Mycobacterium smegmatis 16S rRNA to reconstruct the in-vivo evolution of aminoglycoside resistance in Mycobacterium tuberculosis. Mol Microbiol 77:830–840. https://doi.org/10.1111/j.1365-2958.2010.07218.x

4. Olivares J, Álvarez-Ortega C, Martinez JL (2014) Metabolic compensation of fitness costs associated with overexpression of the multidrug efflux pump MexEF-OprN in Pseudomonas aeruginosa. Antimicrob Agents Chemother 58:3904–3913. https://doi.org/10.1128/AAC.00121-14

5. Olivares J, Alvarez-Ortega C, Linares JF et al (2012) Overproduction of the multidrug efflux pump MexEF-OprN does not impair Pseudomonas aeruginosa fitness in competition tests, but produces specific changes in bacterial regulatory networks. Environ Microbiol 14:1968–1981. https://doi.org/10.1111/j.1462-2920.2012.02727.x

6. Köhler T, van Delden C (2001) Overexpression of the MexEF-OprN multidrug efflux system affects cell-to-cell signaling in Pseudomonas aeruginosa. J Bacteriol 183:5213–5222. https://doi.org/10.1128/JB.183.18.5213

7. Pearson J, Van Delden C, Iglewski B (1999) Active efflux and diffusion are involved in transport of Pseudomonas aeruginosa cell-to-cell signals. J Bacteriol 181:1203–1210

8. Aendekerk S, Diggle SP, Song Z et al (2005) The MexGHI-OpmD multidrug efflux pump controls growth, antibiotic susceptibility and virulence in Pseudomonas aeruginosa via 4-quinolone-dependent cell-to-cell communication. Microbiology 151:1113–1125. https://doi.org/10.1099/mic.0.27631-0

9. Minagawa S, Inami H, Kato T et al (2012) RND type efflux pump system MexAB-OprM of Pseudomonas aeruginosa selects bacterial languages, 3-oxo-acyl-homoserine lactones, for cell-to-cell communication. BMC Microbiol 12:70. https://doi.org/10.1186/1471-2180-12-70

10. Moore JD, Gerdt JP, Eibergen NR, Blackwell HE (2014) Active efflux influences the potency of quorum sensing inhibitors in Pseudomonas aeruginosa. Chembiochem 15:435–442. https://doi.org/10.1002/cbic.201300701

11. Williams P, Cámara M (2009) Quorum sensing and environmental adaptation in Pseudomonas aeruginosa: a tale of regulatory networks and multifunctional signal molecules. Curr Opin Microbiol 12:182–191. https://doi.org/10.1016/j.mib.2009.01.005

12. Withers H, Swift S, Williams P (2001) Quorum sensing as an integral component of gene regulatory networks in gram-negative bacteria. Curr Opin Microbiol 4:186–193. https://doi.org/10.1016/S1369-5274(00)00187-9

13. Swift S, Downie JA, Whitehead NA et al (2001) Quorum sensing as a population-density-dependent determinant of bacterial physiology. Adv Microb Physiol 45:199–270

14. Miller M, Bassler B (2001) Quorum sensing in bacteria. Annu Rev Microbiol 55:165–199. https://doi.org/10.1146/annurev.micro.55.1.165

15. Schuster M, Greenberg EP (2006) A network of networks: quorum-sensing gene regulation in Pseudomonas aeruginosa. Int J Med Microbiol 296:73–81. https://doi.org/10.1016/j.ijmm.2006.01.036

16. Mangwani N, Dash HR, Chauhan A, Das S (2012) Bacterial quorum sensing: functional features and potential applications in biotechnology. J Mol Microbiol Biotechnol 22:215–227. https://doi.org/10.1159/000341847

17. Reading NC, Sperandio V (2006) Quorum sensing: the many languages of bacteria. FEMS Microbiol Lett 254:1–11. https://doi.org/10.1111/j.1574-6968.2005.00001.x

18. Nealson K, Platt T, Hastings J (1970) Cellular control of the synthesis and activity of the bacterial luminescent system. J Bacteriol 104:313–322

19. Nealson K, Hastings J (1979) Bacterial bioluminescence: its control and ecological significance. Microbiol Rev 43:496–518

20. Keller L, Surette MG (2006) Communication in bacteria: an ecological and evolutionary perspective. Nat Rev Microbiol 4:249–258. https://doi.org/10.1038/nrmicro1383

21. Williams P, Winzer K, Chan WC, Cámara M (2007) Look who's talking: communication and quorum sensing in the bacterial world. Philos Trans R Soc Lond Ser B Biol Sci 362:1119–1134. https://doi.org/10.1098/rstb.2007.2039

22. Jayaraman A, Wood TK (2008) Bacterial quorum sensing: signals, circuits, and implications for biofilms and disease. Annu Rev Biomed

Eng 10:145–167. https://doi.org/10.1146/annurev.bioeng.10.061807.160536

23. Martínez JL (2014) Interkingdom signaling and its consequences for human health. Virulence 5:243–244. https://doi.org/10.4161/viru.28073

24. Williams P (2007) Quorum sensing, communication and cross-kingdom signalling in the bacterial world. Microbiology 153:3923–3938. https://doi.org/10.1099/mic.0.2007/012856-0

25. Shaw P, Ping G, Daly S (1997) Detecting and characterizing N-acyl-homoserine lactone signal molecules by thin-layer chromatography. Proc Natl Acad Sci U S A 94:6036–6041

26. Yates EA, Philipp B, Buckley C et al (2002) N-acylhomoserine lactones undergo lactonolysis in a pH-, temperature-, and acyl chain length-dependent manner during growth of *Yersinia pseudotuberculosis* and *Pseudomonas aeruginosa*. Infect Immun 70:5635–5646. https://doi.org/10.1128/IAI.70.10.5635

27. Winson M, Swift S, Fish L (1998) Construction and analysis of luxCDABE-based plasmid sensors for investigating N-acyl homoserine lactone-mediated quorum sensing. FEMS Microbiol Lett 163:185–192

28. Smith RS, Fedyk ER, Springer TA et al (2001) IL-8 production in human lung fibroblasts and epithelial cells activated by the *Pseudomonas* autoinducer N-3-oxododecanoyl homoserine lactone is transcriptionally regulated by NF-kappa B and activator protein-2. J Immunol 167:366–374. https://doi.org/10.4049/jimmunol.167.1.366

29. Zimmermann S, Wagner C, Müller W et al (2006) Induction of neutrophil chemotaxis by the quorum-sensing molecule N-(3-oxododecanoyl)-L-homoserine lactone. Infect Immun 74:5687–5692. https://doi.org/10.1128/IAI.01940-05

30. Wagner C, Zimmermann S, Brenner-Weiss G et al (2007) The quorum-sensing molecule N-3-oxododecanoyl homoserine lactone (3OC12-HSL) enhances the host defence by activating human polymorphonuclear neutrophils (PMN). Anal Bioanal Chem 387:481–487. https://doi.org/10.1007/s00216-006-0698-5

31. Diggle SP, Matthijs S, Wright VJ et al (2007) The *Pseudomonas aeruginosa* 4-quinolone signal molecules HHQ and PQS play multifunctional roles in quorum sensing and iron entrapment. Chem Biol 14:87–96. https://doi.org/10.1016/j.chembiol.2006.11.014

32. Steindler L, Venturi V (2007) Detection of quorum-sensing N-acyl homoserine lactone signal molecules by bacterial biosensors. FEMS Microbiol Lett 266:1–9. https://doi.org/10.1111/j.1574-6968.2006.00501.x

33. Fletcher MP, Diggle SP, S a C et al (2007) A dual biosensor for 2-alkyl-4-quinolone quorum-sensing signal molecules. Environ Microbiol 9:2683–2693. https://doi.org/10.1111/j.1462-2920.2007.01380.x

34. McClean K, Winson M, Fish L (1997) Quorum sensing and *Chromobacterium violaceum*: exploitation of violacein production and inhibition for the detection of N-acylhomoserine lactones. Microbiology 143(Pt 12):3703–3711

35. Diggle SP, Winzer K, Lazdunski A et al (2002) Advancing the quorum in *Pseudomonas aeruginosa*: MvaT and the regulation of N-acylhomoserine lactone production and virulence gene expression. J Bacteriol 184:2576–2586. https://doi.org/10.1128/JB.184.10.2576

36. Winzer K, Falconer C, Garber NC et al (2000) The *Pseudomonas aeruginosa* lectins PA-IL and PA-IIL are controlled by quorum sensing and by RpoS. J Bacteriol 182:6401–6411. https://doi.org/10.1128/JB.182.22.6401-6411.2000

37. Swift S, Karlyshev A, Fish L (1997) Quorum sensing in *Aeromonas hydrophila* and *Aeromonas salmonicida*: identification of the LuxRI homologs AhyRI and AsaRI and their cognate N-acylhomoserine. J Bacteriol 179:5271–5281

Chapter 2

Construction and Use of *Staphylococcus aureus* Strains to Study Within-Host Infection Dynamics

Gareth McVicker, Tomasz K. Prajsnar, and Simon J. Foster

Abstract

The study of the dynamics that occur during the course of a bacterial infection has been attempted using several methods. Here we discuss the construction of a set of antibiotic-resistant, otherwise-isogenic *Staphylococcus aureus* strains that can be used to observe the progress of systemic disease in a mouse model at various time-points postinfection. The strains can likewise be used to study the progression of infection in other animal infection models, such as the zebrafish embryo. Furthermore, the use of antibiotic resistance tags provides a convenient system with which to investigate the effect of antimicrobial chemotherapy during disease.

Key words Infection, Dynamics, Mutant, Isogenic, Animal model

1   Introduction

Research into the problem of antibiotic resistance depends not only upon biochemical, genetic and pharmacological data obtained from bacteria, but also upon studies that investigate the interactions between the pathogen and its host. Posttreatment, antimicrobial compounds accumulate at a range of concentrations within different human organs [1], and bacteria themselves often display a degree of tissue tropism [2, 3]. Even during systemic infections, organs such as the kidneys and liver can act as a large reservoir for bacterial cells [4].

Quantitative attempts to study host-pathogen dynamics during infection rely upon tagging bacterial subpopulations and observing the change in their in vivo distribution over time. This has previously been achieved using signature DNA tags that can be amplified after isolation of the pathogen from the diseased host [5]. An alternative method involves the insertion of various antibiotic resistance cassettes into the bacterial genome at a neutral locus, in order to create a set of selectable, otherwise-isogenic strains [4, 6]. Plasmid-borne antibiotic resistance markers are attractive for

Stephen H. Gillespie (ed.), *Antibiotic Resistance Protocols*, Methods in Molecular Biology, vol. 1736,
https://doi.org/10.1007/978-1-4939-7638-6_2, © Springer Science+Business Media, LLC 2018

their ease of transformation into bacterial cells, but are unfavorable for in vivo experiments due to the propensity for loss of the plasmid when not maintained by selective pressure. Furthermore, differences in plasmid propagation and copy number may introduce unintentional experimental bias. It is therefore preferable to integrate the antibiotic resistance marker into the chromosome.

Whilst experiments are traditionally carried out in the absence of antibiotic treatment of the host organism, one key advantage of using resistance markers for quantification is the additional ability to treat the infected host with the relevant antibiotics in order to study the effect of chemotherapeutic intervention on drug-resistant bacteria. The response of the invading pathogen to subcurative doses, such as those that might be encountered during ineffective or incomplete treatment, is of particular interest. The strains described in this protocol have been successfully used in such experiments [4].

2   Materials

Use ultrapure deionized water (resistivity 18 MΩ cm at 25 °C) during all steps. Unless otherwise stated, store solutions and materials at room temperature. DNA solutions should be stored at −20 °C. Standard molecular cloning techniques and kits should be used; where our methods differ from manufacturers' protocols, the modified methods are explained in full. This protocol assumes basic molecular biology knowledge and equipment/reagents suitable for cloning DNA in *E. coli* only. Molecular biology methods for working with *S. aureus* are given in detail where appropriate. Where filter sterilization is mentioned, use 0.45 μm pore size membranes.

2.1 Strain Construction: Suicide Vectors

1. Oligonucleotide primers (*see* Subheading 3.1 and **Note 1**).

2. Purified genomic DNA (or a fresh colony) of the target organism (*see* **Note 2**).

3. Replication-permissive cells containing a suitable suicide vector (*see* **Note 3**).

4. Preferred high-fidelity PCR reagents.

5. Preferred DNA ligase.

6. Preferred linear DNA and plasmid purification ("miniprep") kits.

7. Agarose and equipment for gel purification of DNA.

8. Standard equipment and reagents for heat shock or electroporation of *E. coli*.

9. Lysogeny broth (LB): 10 g/L tryptone, 5 g/L NaCl, 5 g/L yeast extract. Supplement with 15 g/L agar to make plates. Mix together with water and autoclave, then cool to approximately 45 °C prior to adding appropriate antibiotics.

10. Antibiotics for selection of plasmid-containing cells.

2.2 Strain Construction: Integration of Suicide Vectors into S. aureus RN4220

1. Premixed brain heart infusion (BHI) media made according to manufacturer's instructions. Supplement with 15 g/L agar to make plates. Mix together with water and autoclave, then cool to approximately 55 °C prior to adding appropriate antibiotics.

2. Antibiotics for selection of integrates.

3. Suitable electroporation machine (e.g., Bio-Rad Gene Pulser).

4. Standard laboratory centrifuge (appropriate for 50 mL tubes at ~ 4000 × g).

5. Sterile phosphate buffered saline (PBS); commonly made by dissolving tablets in water then autoclaving.

6. Sterile 10% (v/v) glycerol solution; either autoclaved or filtered.

7. Sterile conical flasks of an appropriate size for growth of 10 mL and 500 mL bacterial culture (at least fourfold larger vessel volume is recommended for efficient aeration).

8. Suitable sterile tubes for bacterial aliquots and centrifugation (1.5 mL, 30 mL, 50 mL).

2.3 Strain Construction: Bacteriophage Transduction of Mutations Between S. aureus Strains

1. Premixed BHI media made according to manufacturer's instructions.

2. LK medium: 10 g/L tryptone, 5 g/L KCl, 5 g/L yeast extract. Supplement with 15 g/L agar and 0.5 g/L sodium citrate to make "phage agar" plates. Mix together with water and autoclave, then cool to approximately 45 °C prior to adding appropriate antibiotics.

3. Phage buffer: 2 mM $MgSO_4$, 5 mM $CaCl_2$, 0.1 M NaCl, 2.5 mM Tris–HCl (pH 7.8), 0.1% (w/v) gelatin. Autoclave to dissolve and sterilize.

4. 1 M $CaCl_2$ solution. Autoclave or filter to sterilize.

5. 20 mM sodium citrate. Autoclave or filter to sterilize.

6. φ11 bacteriophage lysate harvested from a previous preparation.

7. Suitable sterile tubes for transduction and centrifugation (30 mL, 50 mL).

8. Standard laboratory centrifuge (appropriate for 50 mL tubes at ~4000× g).

2.4 In Vivo Studies: Preparation of Bacteria for Mouse Infection

1. Premixed BHI media made according to manufacturer's instructions. Supplement with 15 g/L agar to make plates. Mix together with water and autoclave, then cool to approximately 45 °C prior to adding appropriate antibiotics.

2. Sterile PBS; commonly made by dissolving tablets in water then autoclaving.

3. Dry ice (for snap-freezing samples).

4. Suitable sterile tubes for bacterial aliquots (15 mL).

5. Standard laboratory centrifuge (appropriate for 50 mL tubes at $\sim 4000 \times g$).

6. Standard laboratory vortex mixer.

2.5 In Vivo Studies: Mouse Infection and Quantification of Bacterial Load

1. Premixed BHI media made according to manufacturer's instructions, supplemented with 15 g/L agar to make plates. Mix together with water and autoclave, then cool to approximately 45 °C prior to adding appropriate antibiotics.

2. Sterile PBS; commonly made by dissolving tablets in water then autoclaving.

3. Suitable sterile tubes for mixing bacterial inoculation (7 mL).

4. Suitable sterile needles and syringes for mouse tail vein injection.

5. Sterile tools for extracting organs for quantification of bacterial load.

6. Automatic homogenizer (e.g., Peqlab PreCellys 24-Dual).

7. Sterile 7 mL tubes containing 2.8 mm ceramic beads, compatible with homogenizer.

3 Methods

3.1 Strain Construction: Suicide Vectors

The workflow is summarized in Fig. 1.

For all plating/growth steps, use LB supplemented with antibiotics as indicated.

1. Using standard techniques or software, design oligonucleotide primers to amplify the region of homology from the bacterial chromosome into which you wish to insert your antibiotic cassette (*see* **Note 1**). Include a unique restriction endonuclease recognition sequence at the 5′ end of each primer for insertion of the fragment into the vector.

2. Amplify the DNA fragment with a high-fidelity polymerase and purify it using a commercial kit or similar technique (*see* **Note 2**).

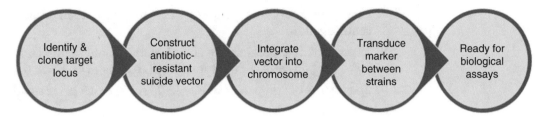

Fig. 1 The workflow presented in this chapter for construction of antibiotic-resistant strains

3. Obtain vector plasmid DNA (*see* **Note 3**) using a commercial mini-prep kit.

4. Digest approximately 1–2 μg of PCR and plasmid DNA and purify the relevant insert and vector fragments using an agarose gel purification kit.

5. Ligate your insert into the vector using a standard ligation protocol. Include a positive (e.g., single-enzyme digested vector) and negative (e.g., water instead of insert DNA) control.

6. Transform ligation mixtures into a high-efficiency *E. coli* cloning strain such as DH5α or TOP10 using a standard heat shock or electroporation protocol. Plate onto appropriate antibiotic agar to select for successful transformants.

7. Recover colonies and verify the sequence of ligated plasmids (*see* **Note 4**).

3.2 Strain Construction: Integration of Suicide Vectors into S. aureus RN4220

For all plating/growth steps, use BHI supplemented with antibiotics as indicated.

1. Grow 10 mL overnight culture of *S. aureus* RN4220 (a common restriction-deficient cloning intermediate) at 37 °C with rapid aeration (e.g., 250 rpm).

2. Subculture into 500 mL prewarmed medium such that $OD_{600} = 0.1$. Grow as above for approximately 1 h until the culture reaches $OD_{600} = 0.4$–0.6 (*see* **Note 5**).

3. Aliquot 4 × 50 mL samples and pellet cells in a centrifuge for 10 min at approximately ~$4000 \times g$). Discard the supernatant then add a further 50 mL culture to each tube and repeat the centrifugation. Discard the second supernatant.

4. Resuspend each pellet in 25 mL sterile, room temperature water. Repeat centrifugation. Discard supernatant. Repeat this step twice more.

5. Resuspend each pellet in 20 mL sterile, room temperature 10% (v/v) glycerol. Repeat centrifugation. Discard supernatant.

6. Resuspend each pellet in 10 mL sterile, room temperature 10% (v/v) glycerol. Combine into a single tube.

7. Leave at room temperature for 30 min, then repeat centrifugation.

8. Resuspend in approximately 400 μL sterile, room temperature 10% (v/v) glycerol. Aliquot into 50 μL volumes (*see* **Note 6**).

9. To aliquots of competent cells, add 0, 5, 10, or 15 μL plasmid mini-prep.

10. Transfer to an electroporation cuvette at room temperature.

11. Electroporate with the following settings: 100 Ω, 2.3 kV, 25 μF.

12. Add 1 mL medium and transfer mixture to a 30 mL universal tube.

13. Incubate for 3 h at 37 °C with rapid aeration (e.g., 250 rpm).

14. Plate onto agar containing antibiotics at 0.2 × the standard concentration (*see* **Note 7**). Incubate for 24–48 h at 37 °C until healthy bacterial colonies appear.

15. Recover colonies and verify integration by colony PCR (*see* **Note 8**). You may also restreak colonies onto the standard antibiotic concentration to verify antibiotic resistance.

3.3 Strain Construction: Bacteriophage Transduction of Mutations Between S. aureus Strains

1. Grow 5 mL overnight culture of the donor strain (RN4220 containing your integrated antibiotic resistance cassette from Subheading 3.2) in BHI at 37 °C with rapid aeration (e.g., 250 rpm).

2. In a 30 mL tube, add 5 mL phage buffer, 5 mL BHI, 150 μL overnight culture, and 100 μL phage lysate from a previous preparation (*see* **Notes 9** and **10**).

3. Incubate the mixture at 25 °C without agitation until it clarifies (indicating complete lysis of bacterial cells) (*see* **Note 11**).

4. Filter-sterilize the lysate. Filtered bacteriophage lysates can be stored for several years at 2–8 °C so long as they remain uncontaminated.

5. Grow a 50 mL overnight culture of the recipient strain in LK at 37 °C with rapid aeration (e.g., 250 rpm).

6. Centrifuge the culture at room temperature for 10 min at approximately ~4000×*g*). Discard the supernatant.

7. Resuspend the pellet in 3 mL LK.

8. Prepare the following in 30 mL tubes:

   (a) 500 μL donor phage lysate, 500 μL recipient culture, 1 mL LK, 10 μL 1 M CaCl₂.

   (b) 500 μL recipient cells, 1.5 mL LK, 15 μL 1 M CaCl₂ (as negative control).

9. Incubate for 25 min at 37 °C without agitation.

10. Incubate for 15 min at 37 °C with rapid aeration (e.g., 250 rpm). Use this time to prechill a centrifuge to 4 °C and place 20 mM sodium citrate on ice.

11. Add 1 mL ice-cold 20 mM sodium citrate to the mixture and incubate on ice for 5 min.

12. Centrifuge the mixture at 4 °C for 10 min at approximately ~$4000 \times g$).

13. Discard the supernatant (*see* **Note 12**).

14. Resuspend the pellet in 1 mL ice-cold 20 mM sodium citrate and incubate on ice for between 45 min and 1.5 h.

15. Spread 100 μL aliquots onto phage agar plates containing the appropriate antibiotic for selection of transductants (*see* **Note 13**).

16. Incubate for 24–48 h at 37 °C until healthy bacterial colonies appear, then restreak and incubate again overnight to ensure complete loss of residual phage particles.

3.4 In Vivo Studies: Preparation of Bacteria for Mouse Infection

1. Grow 5 mL overnight culture of each strain in BHI at 37 °C with rapid aeration (e.g., 250 rpm).

2. Subculture in 50 mL to OD_{600} = 0.01 and grow as above until the culture reaches an appropriate growth phase for your experiment.

3. Stop growth by incubating on ice for 15 min. Use this time to prechill a centrifuge to 4 °C.

4. Vortex the samples well and precisely measure the OD_{600} immediately prior to the next step.

5. Centrifuge the mixture at 4 °C for 10 min at approximately ~$4000 \times g$). Use this time to prechill 15 mL tubes on dry ice.

6. Discard the supernatant, then resuspend the pellet in sterile, ice-cold PBS to approximately 1×10^{10} CFU/mL (*see* **Note 14**). Vortex well.

7. Pipette 100 μL aliquots into 15 mL tubes on dry ice. Allow to freeze.

8. Transfer to −80 °C for long term storage and at least overnight prior to first concentration test. Samples are viable for several years (periodically retest the concentration as described below).

9. Defrost three aliquots of each strain on ice. Add sterile, ice-cold PBS to each tube so that its final concentration is 1×10^8 CFU/mL (*see* **Note 14**).

10. Vortex well, then make a decimal serial dilution series of each aliquot. Plate 100 μL of 10^{-4}, 10^{-5}, and 10^{-6} dilutions onto BHI agar without antibiotics.

11. Incubate plates overnight at 37 °C. This will enable you to detect any nonstaphylococcal contamination and precisely calculate the average final concentration of each strain's aliquots (*see* **Note 15**).

3.5 In Vivo Studies: Mouse Infection and Quantification of Bacterial Load

1. Defrost an aliquot of each strain on ice.

2. Resuspend each aliquot in ice-cold PBS such that its final concentration is 1×10^8 CFU/mL.

3. Mix together an appropriate amount of each aliquot on ice to create a suspension of bacteria at the ratio required by your experiment. Vortex well, both before and after mixing.

4. Prepare and infect mice according to local procedures and experimental plan.

5. Immediately after infection, prepare a decimal serial dilution of the mixed inoculum and plate 100 μL of dilutions 10^{-4}, 10^{-5} and 10^{-6} onto BHI agar plates containing each of the relevant antibiotics individually (*see* **Note 16**). Incubate overnight at 37 °C and record growth to verify the dose and strain ratio used in the infection.

6. After sacrificing animals according to local procedures and experimental plan, extract organs using sterile tools and place into sterile 7 mL homogenizer tubes containing 2.8 mm ceramic beads (*see* **Note 17**). Organs may be frozen at −20 °C until homogenized.

7. Add a known volume of sterile PBS (e.g., 1 mL) to each tube containing an organ. Homogenize according to machine manufacturer's guidelines.

8. Prepare a decimal serial dilution of the organ homogenate and plate 100 μL aliquots of each dilution onto BHI agar containing relevant antibiotics.

9. Incubate overnight at 37 °C and record growth to calculate the bacterial load per strain per organ. Remember to take into account the addition of PBS prior to homogenization.

10. For comparisons between two CFU counts, a two-tailed, unpaired Student's t-test may be used. For comparisons of bacterial strain ratios between two groups (e.g., treated and nontreated), a nonparametric test such as Mann-Whitney U or Kruskall–Wallis is preferable. Common statistical analysis programs such as GraphPad Prism are capable of performing these tests. *See* Fig. 2 for an example of the output from these experiments.

4 Notes

1. Design primers to provide approximately 1 kb of homology to the insertion site. This region will be duplicated in the final merodiploid construct, with the plasmid backbone (and antibiotic resistance cassette) inserted between the two matching regions. Since it is the goal of these experiments to not alter

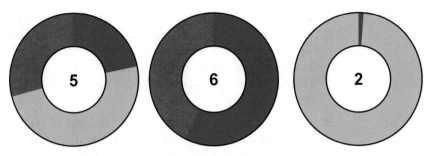

Fig. 2 Output of a three-strain infection study. Each ring shows the strains isolated from a single animal. The number within each ring gives $\log_{10}$(total CFU), i.e., the overall bacterial load. (**a**) Three strains represented approximately equally. (**b**) Two dominant strains; the third is missing. (**c**) One dominant strain, but the low CFU total implies that the viable counts were close to the detection limit. Care should be taken when extrapolating strain ratios from low-CFU data

the host organism except for the introduction of a resistance cassette, consider carefully the orientation of the plasmid-borne genes and their effects on surrounding loci once inserted into the host. For example, it is recommended to integrate the plasmid downstream of an operon that can be easily assayed for disruption (e.g., the terminal gene in an amino acid biosynthesis pathway): once constructed, the *S. aureus* mutants can be screened and any auxotrophs discarded.

2. You may use either purified genomic DNA or a fresh bacterial colony as the template for this amplification. Note that some high-fidelity enzymes cannot tolerate a high concentration of cellular debris, so for colony PCR, dilution of a colony in 100 µL water (then using 1 µL of this as the final reaction template) is recommended.

3. The suicide vector to be used should replicate in *E. coli* but not *S. aureus*, such that it can be manipulated easily but then integrates into the chromosome upon transformation into the latter species. pMUTIN4 is a good example of a compatible suicide vector for integration into gram-positive organisms. Alternatively, it may be possible to use a temperature-sensitive replicon.

4. Colony PCR can be used to analyze transformants. It is recommended that plasmids passing this initial screening step are subsequently sequenced to confirm the absence of point mutations.

5. This step should take approximately 1 h. Check culture turbidity after 30 min and discard if growth takes longer than 90 min.

6. If transforming a suicide vector, the competent cell preparation described should be used immediately. If using this protocol for transformation of a shuttle or temperature-sensitive vector, aliquots may be frozen at −80 °C for long term storage and gently defrosted prior to use.

7. In *S. aureus*, standard antibiotic concentrations are 5 μg/mL tetracycline (Tet marker), 50 μg/mL kanamycin (Kan marker), or 5 μg/mL erythromycin plus 25 μg/mL lincomycin (Ery marker). Note that several steps in the transformation and transduction protocols described herein use altered concentrations.

8. "Junction PCR" with one primer outside of your homology region and a second primer within the plasmid backbone can be used in colony PCR to analyze these clones.

9. The lysate from a previous preparation should ideally contain no selectable markers (i.e., use a lysate grown on the wild type recipient). If a "wild type" donor phage cannot be obtained, instead use a phage grown on a strain with a different resistance marker to the one being selected for in this transduction. These precautions help to prevent the accidental transduction of incorrect mutations.

10. If solution does not appear turbid, add more bacterial culture as required. It is critical that the culture appears slightly cloudy to be able to judge the next step.

11. This normally occurs overnight. If the mixture clarifies within a few hours, add more donor culture until the mixture becomes turbid, then leave overnight.

12. It is important to completely drain all liquid from the pellet at this stage. Ensure this by inverting the tube onto clean tissue paper and tapping gently.

13. Use approximately five plates per lysate and 2–5 plates per control. It is important not to plate too many bacteria on a single plate; if 100 μL produces a lawn of background growth, reduce the volume in future experiments to obtain clean colonies.

14. The approximate concentration of *S. aureus* SH1000 at $OD_{600} = 1$ is 2×10^8 CFU/mL. You may need to adjust this value depending upon your own findings.

15. The concentration of aliquots from a single strain preparation should not vary by more than 10%. If they do, your studies may be affected by imprecise bacterial numbers, and you should consider remaking the aliquots. The most likely cause of concentration variation is failure to properly mix the bacterial suspension prior to pipetting.

16. It is best to perform this plating in duplicate or triplicate so you can obtain an average value for the final strain ratio (and dose) used in the infection.

17. If you intend to homogenize the organs without an automated machine, place the organs in sterile 7 mL tubes and follow the rest of these instructions, simply performing the homogenization step manually.

References

1. Law V, Knox C, Djoumbou Y et al (2013) DrugBank 4.0: shedding new light on drug metabolism. Nucleic Acids Res 42(Database issue):D1091–D1097. https://doi.org/10.1093/nar/gkt1068

2. Horst SA, Hoerr V, Beineke A et al (2012) A novel mouse model of Staphylococcus Aureus chronic osteomyelitis that closely mimics the human infection: an integrated view of disease pathogenesis. Am J Pathol 181:1206–1214

3. Hume EB, Cole N, Khan S et al (2005) A *Staphylococcus aureus* mouse keratitis topical infection model: cytokine balance in different strains of mice. Immunol Cell Biol 83:294–300

4. McVicker G, Prajsnar TK, Williams A et al (2014) Clonal expansion during *Staphylococcus aureus* infection dynamics reveals the effect of antibiotic intervention. PLoS Pathog 10:e1003959

5. Grant AJ, Restif O, McKinley TJ et al (2008) Modelling within-host spatiotemporal dynamics of invasive bacterial disease. PLoS Biol 6:e74

6. Prajsnar TK, Hamilton R, Garcia-Lara J et al (2012) A privileged intraphagocyte niche is responsible for disseminated infection of *Staphylococcus aureus* in a zebrafish model. Cell Microbiol 14:1600–1619

Chapter 3

Method for Detecting and Studying Genome-Wide Mutations in Single Living Cells in Real Time

Marina Elez, Lydia Robert, and Ivan Matic

Abstract

DNA sequencing and fluctuation test have been choice methods for studying DNA mutations for decades. Although invaluable tools allowing many important discoveries on mutations, they are both highly influenced by fitness effects of mutations, and therefore suffer several limits. Fluctuation test is for example limited to mutations that produce an identifiable phenotype, which is the minority of all generated mutations. Genome-wide extrapolations using this method are therefore difficult. DNA sequencing detects almost all DNA mutations in population of cells. However, the obtained population mutation spectrum is biased because of the fitness effects of newly generated mutations. For example, mutations that affect fitness strongly and negatively are underrepresented, while those with a strong positive effect are overrepresented. Single-cell genome sequencing can solve this problem. However, sufficient amount of DNA for this approach is obtained by massive whole-genome amplification, which produces many artifacts.

We describe the first direct method for monitoring genome-wide mutations in living cells independently of their effect on fitness. This method is based on the following three facts. First, DNA replication errors are the major source of DNA mutations. Second, these errors are the target for an evolutionarily conserved mismatch repair (MMR) system, which repairs the vast majority of replication errors. Third, we recently showed that the fluorescently labeled MMR protein MutL forms fluorescent foci on unrepaired replication errors. If not repaired, the new round of DNA synthesis fixes these errors in the genome permanently, i.e., they become mutations. Therefore, visualizing foci of the fluorescently tagged MutL in individual living cells allows detecting mutations as they appear, before the expression of the phenotype.

Key words Mutation, Single-cell, Real-time, Genomic, Microscopy, MutL, Mismatch repair

1 Introduction

Mutations are the raw material of evolution because they are the ultimate source of all genetic variation. Newly arisen mutations can have very different impact on the fitness of the organism, ranging from deleterious through neutral to beneficial. Quantifying when and how different mutations occur allows understanding and predicting the evolution of organisms. In bacteria, these parameters are for example useful for predicting the rapidity with which they can evolve different capacities, such as antibiotic resistance and

Stephen H. Gillespie (ed.), *Antibiotic Resistance Protocols*, Methods in Molecular Biology, vol. 1736,
https://doi.org/10.1007/978-1-4939-7638-6_3, © Springer Science+Business Media, LLC 2018

evasion of the host immune system. In addition, many antibiotics are known to increase mutations rates in bacteria, which can increase the evolvability of bacterial pathogens.

DNA mutations have been studied for more than a century but never directly observed. Initially, the occurrence of mutations in genomes of organisms was inferred from the detection of mutational events in genetic markers using selection based on gain or loss of a function. Fluctuation test, one of the first quantitative tests for measuring mutations is based on this property [1, 2]. The fraction of the new phenotypic variants in a growing population is measured and the number of cycles of cell division estimated. This allows for estimating mutation rates. However, these estimates are biased due to the fitness effects of mutations. For example, variants carrying deleterious mutations are often affected for growth and have longer generation times compared to the nonmutated cells. As a consequence, they are outgrown in the final population by the nonmutated cells that divide more rapidly. The rates of variants with deleterious mutations obtained by fluctuation analysis are therefore underestimated. In addition, fluctuation analysis is applicable to minority of DNA mutations, which produce an identifiable phenotype, i.e., lethal, synonymous, and some deleterious mutations cannot be detected. Extrapolating to whole genomes from data obtained from few loci is likely to be inaccurate because mutation rates vary between different chromosomal sites due to the differences in base composition, transcriptional activity and variations in DNA repair efficiency [3].

More straightforward strategy for identifying and quantifying mutations is DNA sequencing. DNA sequencing gives precise information on the mutation site and the mutation type of the vast majority of DNA mutations. It became very rapid and inexpensive. For example, next-generation DNA sequencing generates several billions of nucleotides of DNA sequence in less than 2 weeks [4]. However, 1% of sequenced bases are identified incorrectly due to errors introduced during sample preparation, DNA amplification, and image analysis [4]. This is problematic for detecting the rare mutations and those with strong negative effect on the fitness when sequenced bacterial populations are heterogeneous, which is generally the case. Single-cell genome sequencing can solve this problem. However, sufficient amount of DNA for this approach is obtained by massive whole-genome amplification, which produces many artifacts. Finally, because cells must be killed to extract DNA, it is impossible to perform single-cell time-course analysis using DNA sequencing methods.

The vast majority of spontaneous mutations are due to the DNA polymerase errors occurring during DNA replication. While the error rates of replicative DNA polymerases are of the order of

10^{-4}–10^{-5} per nucleotide, DNA repair pathways reduce replicative error rates to 10^{-9}–10^{-11} per nucleotide [5, 6]. Major contributor to the DNA replication fidelity is mismatch repair (MMR), which eliminates 99.99% of the errors generated by the replicative DNA polymerases [6]. During replication, MMR detects DNA replication errors and recruits enzymes to destroy the portion of the DNA strand that contains the error. MMR is found in all domains of life and its function is evolutionarily highly conserved. To correct errors, MMR protein MutS binds to the sites of DNA replication errors and recruits MutL MMR protein. In enterobacteria, the MutL bound to mismatch-MutS complex recruits MutH, an endonuclease that cleaves the newly replicated DNA strand in the proximity of the error. This triggers the removal of a segment of single-stranded DNA containing the wrong base by the UvrD DNA helicase and ssDNA exonucleases. The repair process is finalized by DNA polymerase III and DNA ligase activity. Eukaryotes use homologs of MutS and MutL to correct errors in DNA replication, but lack a homolog of MutH [6].

We showed that MMR system can be exploited to visualize the unrepaired DNA replication errors, i.e., emerging DNA mutations. By imaging the fluorescently labeled components of MMR, we found that fluorescent MutL protein forms foci on such DNA sites [7], probably due to extensive accumulation of MutL proteins when repair cannot be completed. Our conclusion that the fluorescent MutL foci tag emerging DNA mutations is based on two findings. Previously, DNA sequencing data and fluctuation analysis estimated the mutation rate of wild-type *Escherichia coli* to about one mutation per 300 cells [8, 9]. In good agreement with this, we found that the frequency of fluorescent MutL foci in wild-type *E. coli* cells is about 1 MutL foci per 230 cells [7] (Fig. 1a). Second, we found that cells that mutate 50–1000-fold more compared to wild-type *E. coli* cells show 50–1000-fold more MutL foci [7] (Fig. 1b). We also showed that the mutation rates estimated by fluctuation analyses are proportional to the MutL foci frequency over a several hundred-fold range [7]. This real-time method for mutation detection allows recording mutation rates of thousands of individual cells in less than an hour. Furthermore, using this method we could detect up to seven mutations per single cells as they appear, before the expression of the mutation phenotype. Finally, because MutL foci disappear from the sites of emerging mutations when a new replication fork passes through, detecting fluorescent MutL foci allows direct measuring of the per generation mutation rate. This finding was based on our data showing that preventing new rounds of DNA replication by treating cells with rifampicin prevents disappearance of the MutL foci.

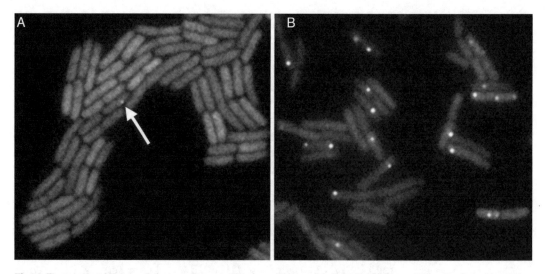

Fig. 1 Fluorescent MutL protein tags the emerging mutations. (**a**) *E. coli* wild-type cells growing on agarose pad and expressing the fluorescent MutL. All cells show uniform cytoplasmic fluorescence. The bright fluorescent MutL spot (indicated by an arrow) tagging emerging mutation is visible in only one cell. (**b**) Population of the *mutH* cells, which are MMR deficient mutants, produce 50–100-fold more mutations and produce more MutL fluorescent foci per cell, than wild-type cells

2   Materials

2.1   Escherichia coli Strains

1. *E. coli* MG1655 expressing the *xfp-mutL* gene and deleted for the native chromosomal nonfluorescent *mutL* gene. The deletion of the native *mutL* gene prevents the recruitment of the nonfluorescent MutL protein to the sites of DNA mutations. This is important for the proper visualization of DNA mutations as the recruitment of the nonfluorescent MutL to such DNA sites could decrease the fluorescent signal of MutL foci or obstruct it completely. Different fluorescent proteins can be used to render the MutL protein fluorescent. We constructed the fusions of genes coding for enhanced green fluorescent protein (*egfp*), yellow fluorescent protein *(yfp)*, or cyan fluorescent protein *(cfp)* to the *mutL* gene. We used these different fluorescent constructs in different experimental setups depending on our need to visualize only the fluorescent MutL, to investigate the colocalization of different MMR proteins [7, 10], or to investigate the colocalization of the fluorescent MutL with the replisome or different regions of the *E. coli* chromosome. *xfp-mutL* gene was either cloned on a plasmid or inserted into *E. coli* chromosome. Chromosomal construction is preferred because it produces less cell-to-cell heterogeneity in the amount of the cytoplasmic fluorescent MutL. Reduction of such noise makes the data analysis easier. When cell-to-cell variations in cytoplasmic fluorescence are limited, the foci can

be detected automatically using a simple thresholding method. We made two different chromosomal *yfp-mutL* constructs. In one case, we inserted *yfp-mutL* at the position of the native *lacZ* such that *yfp-mutL* is expressed from the inducible *lac* promoter (Plac). In the other case, we inserted *yfp-mutL* at the site of native *mutL* gene. In this case *yfp-mutL* is constitutively expressed from the native *mutL* promoter (PmutL). These two constructs are suitable for different applications. Working with fluorescently tagged proteins expressed at their physiological levels might seem preferable as it prevents the toxicity and the nonspecificity of the overproduction of the fluorescently labeled proteins. However, we found that the overexpression of the fluorescent MutL is not toxic and that it does not affect the frequency of MutL foci. In addition, the PmutL is a weak promoter, leading to the synthesis of on average 113 MutL dimers per *E. coli* cell [11], which in the case of the *yfp-mutL* means a low MutL fluorescence. Detection of the low fluorescence requires high excitation light intensity or long exposure of cells to excitation light, which causes the bleaching of fluorescence, leading to the signal loss and underestimations (*see* **Note 1**). Expression of the *yfp-mutL* from the Plac promoter is higher compared to the expression of the *yfp-mutL* from the PmutL. Therefore, detecting fluorescent MutL foci using this construct requires less illumination.

The expression of the *yfp-mutL* from the Plac promoter varies depending on the growth medium. In some media, leaky expression from the Plac will produce enough fluorescent MutL allowing for the complete complementation of the native *mutL* inactivation. In others, the Plac inducer, IPTG, should be added to the growth medium to assure the expression of sufficient amount of fluorescent MutL. It is important to determine before starting a new experiment if IPTG should be supplied to the growth medium or not (*see* **Note 2**). Expression level of the *yfp-mutL* gene should be sufficient to restore wild-type mutation rate to the strain whose native chromosomal *mutL* gene is deleted. This can be done quantitatively or qualitatively (*see* **Note 3**) by a classical mutagenesis experiments. For the quantitative test, grow the strain deleted for the native chromosomal *mutL* gene, which expresses the fluorescent MutL overnight in a desired medium supplemented or not with IPTG. Do the same for the reference wild-type *E. coli* strain. Upon growth to saturation, dilute 10^7-fold the saturated cultures to eliminate preexisting rifampicin-resistant (Rif^R) mutants and grow them again to saturation. Plate the dilutions of the saturated cultures on the selective medium plates, which contains LB supplemented with 100 μg/mL rifampicin to select Rif^R mutants, and on the LB medium plates to determine the total number of colony forming units.

Colonies should be scored after 24 h of incubation at 37 °C. Determine the average frequency of Rif^R mutants from three to six independent experiments. The Rif^R mutant frequency of the strain expressing fluorescent *mutL* and deleted for the native chromosomal *mutL* gene should not be different from that of the nonmodified wild-type reference strain (*see* **Note 4**).

The *yfp-mutL* fusion that we designed and constructed can be transferred into a desirable *E. coli* strain by a classical P1 transduction [12]. This is possible as we cloned the selectable chloramphenicol resistant marker (*cam*) downstream of the *yfp-mutL* gene (*see* **Note 5**). Otherwise, the DNA coding for the *yfp-mutL-cam* can be amplified by PCR, using the plasmid or the chromosomal templates, or synthesized de novo. Obtained DNA fragments can be inserted into a desired position on the *E. coli* chromosome by the Datsenko and Wanner gene replacement method [13] (*see* **Note 6**).

2. *E. coli* strain expressing the *mutL-yfp* and deleted for the native chromosomal *mutL* and *mutS* genes. To check that *mutS* is properly deleted in the strain expressing fluorescent MutL and inactivated for the native chromosomal *mutL* do the qualitative Rif spot test as described above. Because MutS protein is necessary for the MutL protein binding to emerging mutations, this control strain allows distinguishing the "nonfunctional" aggregates of MutL protein, which are MutS independent, from the "functional" MutL foci tagging emerging mutations, which are MutS dependent. Therefore, the presence of fluorescent foci in this strain indicates that the culture conditions lead to nonfunctional fluorescent MutL aggregates and are consequently not suitable to detect mutations by our method.

2.2 Growth Medium

In principle, any growth medium can be used. We used the Plac and the PmutL constructs grown in LB supplemented with 0.1 mM IPTG as well as in standard M9 minimal medium [12] supplemented by 2 mM MgSO$_4$, 0.003% vitamin B1, 0.001% uracil, 0.2% casamino acids, and 0.01% glycerol. If Plac construct should be used in a different growth medium check before starting the experiment the fluorescent level of cells because too much expression from the Plac inducible promoter in minimal media complemented with pyruvate or glycerol could prevent accurate detection of fluorescent MutL foci due to the high background cytoplasmic fluorescence (*see* above). If not using the inducer, check also that enough fluorescent MutL is produced in cells to complement the inactivation of the native chromosomal *mutL* gene (*see* above).

2.3 Microscopy and Mounting

1. Classical epifluorescence microscope equipped with a 100× objective, appropriate fluorescent filters, fluorescent lamp, and CCD, EMCCD, or sCMOS camera.

2. Microscope glass slides and coverslips, Gene Frame and agarose for preparing agar pads.

3 Methods

3.1 Growth

1. Starting from the glycerol stock, grow an overnight culture of the strain expressing *xfp-mutL* in a desired medium to saturation at 37 °C. Add the inducer to the growth medium when necessary (*see* above).

2. Dilute 400-fold the saturated culture and incubate at 37 °C until O.D. 0.15–2.0 is reached (*see* **Notes 7** and **8**).

3. Centrifuge 1 mL of the exponentially growing culture for 1 min at 13,000 rpm (12000 × *g*) to concentrate cells.

4. Throw away the supernatant, resuspend cells by pipetting and depose 1–2 μL of suspension on the agarose pads prepared before (*see* below).

3.2 Preparing Agar Pads

1. Dissolve agarose (1.5%) in the medium in which cells have been grown using a microwave oven (*see* **Note 9**). If the autofluorescence of the growth medium is too high, as in the case of LB, dissolve agarose in the minimal medium or simply in M9 (*see* **Note 10**). If required, supplement the growth medium with the Plac inducer IPTG.

2. Take a clean microscope glass slide (dimensions adapted to your microscope). Attach the Gene Frame, the adhesive system for easy agarose pad preparation, in the middle of the slide (*see* **Note 11**).

3. Transfer 100 μL or 200 μL of the warm agarose in the middle of the Gene Frame, depending on the Gene Frame dimensions, and cover rapidly with the proper coverslip (*see* **Notes 12** and **13**).

4. Leave the slide in a horizontal position for 10 min at the room temperature to allow the agarose to solidify.

5. Remove the coverslip and the Gene Frame upper plastic cover. Plate 1–2 μL of the concentrated exponential phase culture on the agarose pad and allow the liquid to disperse by turning the slide in different directions 3–4 times.

6. Leave to dry for a few minutes at room temperature until no more liquid is detectable on the agarose pad (*see* **Notes 14** and **15**).

7. Put the coverslip, and gently press to assure the proper sealing of the coverslip to Gene Frame. Try avoiding making air bubbles (*see* **Note 13**).

3.3 Microscopy

1. Mount the prepared slide on the epi-fluorescence microscope.

2. Choose the fields with a monolayer of 100–500 cells per field (*see* **Note 16**).

3. Record images at 100× magnification in phase contrast and in fluorescence. The excitation light intensity and the exposure time allowing detecting all fluorescent MutL foci are setup dependent. Each experimenter needs to determine these parameters for his system. Choose the minimum excitation light intensity and the shortest exposure times for which all fluorescent MutL foci are detected. This will decrease phototoxicity and limit fluorescence bleaching. The detection of the fluorescent MutL foci will therefore be more accurate.

4 Notes

1. In addition to bleaching the fluorescent signal, high levels of excitation light are toxic to cells. While this is not the problem when taking a single snapshot picture of growing cells, it is relevant when performing time series imaging. Therefore, we use a low level of excitation light when doing the time-lapse imaging of the microcolonies growing on the agar pads or when doing long-term imaging of cells growing in the microfluidic chips. In these experimental setups, it is necessary to determine experimentally the maximum excitation light level that does not cause cell toxicity when applied with a desired interval. This can be done by comparing, at the end of the experiment, the fitness of the cells that were subjected to the imaging to the fitness of the cells growing in the same setup, but not imaged throughout the experiment.

2. We do not recommend adding the IPTG systematically to the growth medium. In some media, the leaky expression is sufficient and inducing the expression of *yfp-mutL* more will lead to increased background cytoplasmic fluorescence. Too high cytoplasmic fluorescence prevents the accurate detections of fluorescent MutL foci. We found that adding the IPTG inducer is not necessary when cells carrying Plac construct are grown in minimal medium supplemented with casamino acids at 0.2%. On the contrary, in LB, the leaky expression of the fluorescent *mutL* from the Plac is not sufficient for complementing the inactivation of the native *mutL*. The inducer IPTG needs to be added to the LB growth medium at the concentration of 0.1 mM.

3. For the qualitative mutagenesis test (Rif spot test) grow the strain expressing the fluorescent MutL and deleted for the native chromosomal *mutL* gene, as well as the reference wild-

type strain, overnight in a desired medium, supplemented or not with IPTG. Upon growth to saturation, plate 20 μL of the saturated cultures on the LB plates containing 100 μg/mL rifampicin to select Rif[R] mutants and incubate the plates for 24 h at 37 °C. If fluorescent MutL is sufficiently expressed to complement native *mutL* gene inactivation, a small number (<5) of Rif[R] colonies will grow per spot, which is comparable to the wild-type strain. In contrast, if there are significantly (50–100-fold) more colonies per spot than for the wild-type strain, native *mutL* gene inactivation is not fully complemented.

4. For wild-type *E. coli* cells, the expected frequency of Rif[R] mutants is around 2×10^{-8}. Cells inactivated for MMR (*mutS, mutL, mutH, uvrD*) show 50–100-fold higher frequency of Rif[R] mutants.

5. In the case of the Plac construct, the insertion of the *yfp-mutL-cam* in the chromosome leads to the deletion of *lacZ* gene. Thus, white, blue screen can be used to search for positive clones amongst the selected chloramphenicol resistant ones.

6. In this case, make sure that the native chromosomal *mutL* is deleted, not simply inactivated, before transforming the cells with the DNA coding for the *yfp-mutL-cam*. Otherwise, the *yfp-mutL-cam* DNA inserts incompletely in the native chromosomal *mutL* site on the *E. coli* chromosome.

7. If you wish to compare the results obtained in different experiments, pay attention to O.D, because the mutation rate could vary at different O.D.s. Our preliminary results indicate that mutation rates might be lower at the entry into stationary phase compared to mid- and early-exponential growth phase.

8. It is important to dilute the saturated culture at least 400-fold. This allows the majority of cells to exit the stationary phase before imaging. Some stationary phase cells accumulate occasionally "nonfunctional" MutL aggregates, which are not tagging DNA mutations. These foci, contrary to "functional" fluorescent MutL foci, form independently of the MutS protein. Consequently, they are detectable in the stationary phase *mutS* cells. Diluting enough helps getting rid of such nonspecific MutL aggregates.

9. Make sure that the growth medium has been filtered and that agarose has been completely dissolved. This will help to decrease the autofluorescence of the agarose pad.

10. In this case, make sure not to leave the cells on the slide for more than 20 min before imaging. Due to nutrient lack, cells might enter the stationary phase and cease replicating. In these conditions cells mutate less and "nonfunctional" *mutL* aggregates start appearing (*see* **Note 8**).

11. Follow the manufacturer's instructions for the proper sealing of the Gene Frame on the slide. Do not remove the upper plastic cover at this step.

12. This step should be done quickly. The best is to hold the coverslip in one hand while pipetting the agarose with the other hand. This allows putting the coverslip rapidly on the warm agarose, preventing it from solidifying and making the agarose pad very flat. This will assure the unique microscope focus across the entire microscope field. Therefore, all imaged cells will be exploitable for analysis.

13. Try avoiding the air bubbles while pipetting the agarose as the air bubbles have high autofluorescence.

14. Make sure that the agarose pad dries enough after plating the cells on it and before putting the coverslip. White traces of the liquid become visible on the agarose pad that is ready for imaging. Otherwise, if the agarose pad is not dry enough, the cells will not attach to it, they will swim, and the imaging will be impossible.

15. Make sure that the agarose pad does not dry too much and become wrinkled. This will prevent the proper sealing of the coverslip to the agarose pad and degrade the image quality.

16. Avoid the fields with cells in double layer and also fields where too many cells are sticking to each other. In such cases the foci detection will be difficult.

Acknowledgments

This work was supported by the ANR young researcher grant ANR-14-CE09-0015-01 to M.E., and by Idex ANR-11-IDEX-0005-01 / ANR-11-LABX-0071 and AXA Research grants to I. M.

References

1. Luria SE, Delbruck M (1943) Mutations of bacteria from virus sensitivity to virus resistance. Genetics 28:491–511

2. Foster PL (2006) Methods for determining spontaneous mutation rates. Methods Enzymol 409:195–213

3. Nishant KT, Singh ND, Alani E (2009) Genomic mutation rates: what high-throughput methods can tell us. Bioessays 31:912–920

4. Shendure J, Ji H (2008) Next-generation DNA sequencing. Nat Biotechnol 26:1135–1145

5. Kunkel TA (2004) DNA replication fidelity. J Biol Chem 279:16895–16898

6. Li GM (2008) Mechanisms and functions of DNA mismatch repair. Cell Res 18:85–98

7. Elez M, Murray AW, Bi LJ, Zhang XE, Matic I, Radman M (2010) Seeing mutations in single cells. Curr Biol 20:1432–1437

8. Drake JW (1991) A constant rate of spontaneous mutation in DNA-based microbes. Proc Nat Acad Sci USA 88:7160–7164

9. Lee H, Popodi E, Tang H, Foster PL (2012) Rate and molecular spectrum of spontaneous mutations in the bacterium Escherichia Coli as determined by whole-genome sequencing. Proc Natl Acad Sci U S A 109:E2774–E2783

10. Elez M, Radman M, Matic I (2012) Stoichiometry of MutS and MutL at unrepaired mismatches *in vivo* suggests a mechanism of repair. Nucleic Acids Res 40:3929–3938

11. Feng G, Tsui HC, Winkler ME (1996) Depletion of the cellular amounts of the MutS and MutH methyl-directed mismatch repair proteins in stationary-phase *Escherichia coli* K-12 cells. J Bacteriol 178:2388–2396

12. Miller JH (1992) A short course in bacterial genetics. Cold Spring Harbor Press, Cold Spring Harbor, NY

13. Datsenko KA, Wanner BL (2000) One-step inactivation of chromosomal genes in *Escherichia coli* K-12 using PCR products. Proc Natl Acad Sci U S A 97:6640–6645

Chapter 4

Detecting Phenotypically Resistant *Mycobacterium tuberculosis* Using Wavelength Modulated Raman Spectroscopy

Vincent O. Baron, Mingzhou Chen, Simon O. Clark, Ann Williams, Kishan Dholakia, and Stephen H. Gillespie

Abstract

Raman spectroscopy is a non-destructive and label-free technique. Wavelength modulated Raman (WMR) spectroscopy was applied to investigate *Mycobacterium tuberculosis* cell state, lipid rich (LR) and lipid poor (LP). Compared to LP cells, LR cells can be up to 40 times more resistant to key antibiotic regimens. Using this methodology single lipid rich (LR) from lipid poor (LP) bacteria can be differentiated with both high sensitivity and specificity. It can also be used to investigate experimentally infected frozen tissue sections where both cell types can be differentiated. This methodology could be utilized to study the phenotype of mycobacterial cells in other tissues.

Key words Raman spectroscopy, Mycobacteria, Phenotypic resistance, Lipids

1 Introduction

Tuberculosis is a major health issue worldwide and a major cause of death due to infectious disease. Treatment of tuberculosis has not improved in the past 50 years. Shortening therapy would make an important step forward to reducing the global burden of tuberculosis. Recent clinical trials using more bactericidal regimens to shorten TB therapy to 4 months failed to do so due to higher relapse rate after successful treatment [1–3]. Those observations confirmed that relapse is the main barrier to shorter tuberculosis treatment. The bacteriology of relapse remains largely unknown and due to its importance represents a key research area in tuberculosis. Patients that clear *Mycobacterium tuberculosis* from their sputum rapidly during treatment can still undergo relapse [4]. In order to improve our knowledge of relapse and its bacteriology, we need nondestructive methods to study bacteria directly at the site of the disease. Relapse could be linked to bacteria that survive

Stephen H. Gillespie (ed.), *Antibiotic Resistance Protocols*, Methods in Molecular Biology, vol. 1736,
https://doi.org/10.1007/978-1-4939-7638-6_4, © Springer Science+Business Media, LLC 2018

therapy, and several studies have shown that mycobacteria accumulate lipid in intracellular bodies and these cells exhibit a lower metabolic rate [5–8]. Recently lipid body positive mycobacteria were shown to be much more resistant to key components of the tuberculosis therapy, up to 40 times more resistant to rifampicin [9]. Both phenotypes, lipid rich (LR) and lipid poor (LP), can be observed in any mycobacterial population in a range of species. To study the lipid content of *Mycobacterium tuberculosis,* we describe a novel method consisting of an all-optical label-free Raman spectroscopy based system that can be applied to bacteria directly in tissue.

Raman spectroscopy has been applied previously to discriminate cultured bacteria and mycobacteria species but never discriminate between two phenotypes or target mycobacteria in tissue [10–12]. We use wavelength modulated Raman (WMR) spectroscopy to improve both sensitivity and specificity. Rather than using a single excitation wavelength, WMR spectroscopy scans over a small range of the laser wavelengths. Combined with subsequent multivariate statistical analysis, all background fluorescence from biological samples can be removed. Importantly, WMR spectroscopy is a label-free technology and can be therefore combined with other techniques such as immunostaining.

2 Materials

2.1 In-Vitro Investigation

1. Test organism in this case: *Mycobacterium tuberculosis* (NCTC7416).

2. Growth medium: Middlebrook 7H9 medium (Sigma-Aldrich, UK).

3. Glycerol (Sigma-Aldrich, UK).

4. Tween 80 (Fisher BioReagents, UK).

5. Middlebrook ADC enrichment (Sigma-Aldrich, UK).

6. Bacterial culture tube.

7. Incubator set at 37 °C.

8. Phosphate-buffered saline (PBS).

9. Quartz coverslip (01015T-AB, SPI Supplies, PA, USA).

10. Quartz slide (01016-AB, SPI Supplies, PA, USA).

11. Transparent nail polish.

2.2 Tissue Investigation

1. Frozen infected tissue to investigate: in this example infected guinea pig (Specific pathogen-free Dunkin Hartley strain guinea pigs) lung sections.

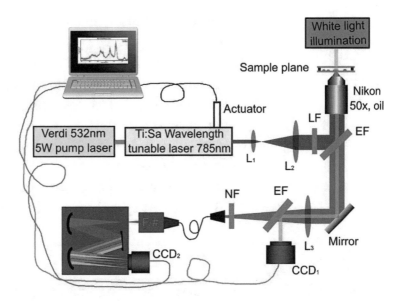

Fig. 1 Schematic diagram of the experimental setup. The system is using a tunable Ti:Sapphire laser (Spectra-Physics 3900 s, 785 nm, 1 W) pumped by a green laser (Verdi V6, 532 nm, 5 W). L_1, L_2, and L_3 are lens. LF denotes line filter. EF denotes edge filter and NF denotes notch filter. The laser is focused on the sample using a microscope objective (Nikon, 50×, oil). The spectrometer is composed of a monochromator (Andor Shamrock SR303i, 400 lines/mm grating at 850 nm) and a cooled CCD camera (Andor Newton)

2. Cryostat to produce 3–5 μm thick frozen tissue sections (*see* **Note 1**).

3. Formalin solution.

4. Sucrose.

5. PBS.

6. Quartz coverslip (01015T-AB, SPI Supplies, PA, USA).

7. Quartz slide (01016-AB, SPI Supplies, PA, USA).

8. Transparent nail polish.

2.3 Raman System   An example of the Raman system design is presented in Fig. 1.

1. Tunable laser that can generate light at 785 nm (*see* **Note 2**).

2. Microscope objective (Nikon, 50×, oil immersion).

3. Spectrometer: formed with a monochromator (Andor Shamrock SR303i, 400 lines/mm grating at 850 nm) and a cooled CCD camera (Newton 920, Andor Technology, UK).

4. A computer linked to the Ramen system and connected to the Internet.

5. Software list: Matlab, Andor Solis.

List of optics:

6. **LF**: Laser line filter (LL01-785, Semrock, USA).

7. **EF**: Edge filter (LPD02-785RU, Semrock, USA).

8. **NF**: Notch filter (NF03-785E, Semrock, USA).

9. **FM**: flip mirror.

10. **CCD$_1$**: Imaging Source USB camera (DFK 42AUC03, Imaging Source, Germany).

11. **CCD$_2$**: Andor Newton Camera (cooled at −70 °C).

12. **L1–L3**: lenses; **F/#**: F number matcher.

3 Methods

3.1 Bacteria from Culture

3.1.1 Culture

Perform all steps before the heat killing of bacilli in a level 3 laboratory facility (or equivalent). Grow *M. tuberculosis* (NCTC7416) at 37 °C in Middlebrook 7H9 medium supplemented with 0.05% (v/v) tween 80 and 2 mL of glycerol (for 450 mL of medium). Add one vial of Middlebrook ADC supplement (M0678, FLUKA) to the 450 mL.

3.1.2 Heat Killing of Bacilli

Harvest 1 mL of *M. tuberculosis* culture and place it at 80 °C for 20 min to heat inactivate the bacteria. Take the inactivated bacteria out of the Cat3 facility (*see* **Note 3**).

3.1.3 Wash

Take 100 μL of bacterial suspension and spin it at 20,000 × *g* for 3 min at room temperature. Discard the supernatant. Wash the cells twice with 100 μL of PBS.

3.1.4 Raman Slide Preparation from In-Vitro Cell

Resuspend the bacteria in 20 μL of PBS, take 10 μL out and heat fix onto a quartz coverslip (SPI Supplies). Mount the fixed bacteria that are on the coverslip, on top of a quartz slide (SPI Supplies). The cells end up between the quartz slides and coverslip. Seal mount using a transparent nail polish (leave to air-dry for an hour before use). Interrogate the sample with WMR spectroscopy.

3.2 Tissue Sample Preparation

3.2.1 Tissue Sectioning

Treat the tissue to investigate with 10% neutral buffered formalin for an appropriate amount of time if it is infected with *M. tuberculosis* (or another pathogen). Remove the tissue from the formalin solution and freeze the tissue sample on a bed of dry ice in OCT (optimal cutting temperature) solution (30% sucrose in PBS) for future Raman investigation. From the OCT block cut 3–5 μm sections using a cryostat. Place the tissue section on the quartz coverslip.

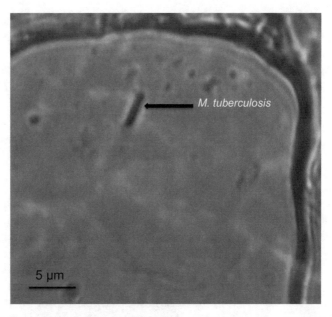

Fig. 2 *M. tuberculosis* in infected guinea pig lung tissue section observed under the Raman system (bright field). A single bacillus can be observed (black arrow) in alveoli. The scale bar represents 5 μm

3.2.2 Tissue Section Mounting

Mount the frozen tissue section that is on the quartz coverslip (SPI Supplies, 01015T-AB), directly on top of a quartz slide (SPI Supplies, 01016-AB). Seal the mount with transparent nail polish (leave to air-dry for an hour before use). Interrogate the tissue sample with WMR spectroscopy. An example of single *M. tuberculosis* bacillus in guinea pig lung tissue section observed under the Raman system is shown in Fig. 2 (*see* **Note 4**).

3.3 Raman Microscopy Methods

3.3.1 Raman System

Use your Raman system to acquire the spectra, an example of a confocal Raman system shown in Fig. 1. This system uses a green laser (Verdi V6, 532 nm, 5 W) to pump a tunable Ti:Sapphire laser (Spectra-Physics 3900 s, 785 nm, 1 W). To focus the laser on the bacteria use an oil immersion objective a Nikon, 50×, oil for example. A spectrometer with a monochromator (Andor Shamrock SR303i, 400 lines/mm grating at 850 nm) with a cooled CCD camera (Andor Newton) is used to collect the Raman photons. Determine the laser power to use according to the sample type being investigated (*see* **Note 5**).

3.3.2 Acquisition

Standard Raman Spectra

Use for each single bacteria a 30 s integration time with a stable excitation laser wavelength at 784.6 nm. Record a separate background Raman spectrum with the same condition from a (bacteria-free position) position near the bacteria. Use it to subtract the background signals afterward. Determine the integration time you need to obtain an optimal signal-to-noise ratio; this could vary depending on your system, the organism under investigation and the power on the sample plane (*see* **Note 6**).

WMR Spectra

Record five spectra continuously, with 30 s of integration each, while the laser is tuning over a small range of 1 nm. From these five original spectra calculate a single WMR spectrum with the autofluorescence background removed. Compare to the standard Raman spectrum; all Raman peaks will locate at the zero crossings while their peak intensity corresponds to the peak-to-valley value (*see* **Note 7**).

3.3.3 Raman Calibration and Spectra Processing

Use control Raman spectra from polystyrene beads (1 μm in diameter) to monitor any possible drift in the laser wavelength or the optical system. Acquire the Raman spectra of the polystyrene beads with the same integration time as the experimental conditions.

The laser wavelength may vary during the experiments. We use a standard chemical, polystyrene, to monitor this variation. As the known largest peak position of polystyrene is at 1001.4 cm^{-1}, the actual laser wavelength can be calculated (*see* **Note 8**).

If the drift in laser wavelength is very small (typically <0.2 nm over a day) and slow, the actual laser line used to acquire each Raman spectrum can be calculated using an interpolation.

To avoid any influence from laser power fluctuation during wavelength tuning, normalize each Raman spectrum by its total intensity (i.e., the integration over the area covered by the spectrum). To compare your data sets and do the data processing use mainly the fingerprint region from 1000 cm^{-1} to 1800 cm^{-1} (*see* **Note 9**).

After Raman investigation, the tissue sample can be used for another method (*see* **Note 10**).

3.3.4 Principal Component Analysis

To distinguish between two different cell phenotypes or species, apply principal component analysis (PCA) to each training dataset containing standard Raman spectra or WMR spectra. Use approximately 60–80 cells for each phenotype or specie.

Use a number of principal components (PCs) that corresponds to more than 70% of variances in the training dataset. In this example, the first seven principal components (PCs) have been used (*see* **Note 11**).

3.3.5 Leave-One-Out Cross Validation

Use the method of leave-one-out cross validation (LOOCV) to estimate the ability of classification for the different cell phenotypes or species. Without considering one Raman spectrum, a multiple-dimensional space is defined by all the PCs in the training data set. This leave-out spectrum is then classified to be a spectrum taken from either of your cell types based on its position in the multiple-dimensional space. With this LOOCV for each spectrum in the data set, the specificity and sensitivity for a data set containing two cell types are calculated (*see* **Note 11**).

4   Notes

1. Frozen tissue slicing

   We have used 5 μm thick formalin-fixed, frozen tissue sections successfully. However, the tissue thickness can be adjusted according to the experiment. Commercial companies can slice fixed, frozen tissue and mount onto quartz slides if cryostat equipment is not available.

2. Laser

   A laser that can generate light at 785 nm and its wavelength can be tuned over 1 nm. For example, a tunable Ti:Sapphire laser (Spectra-Physics 3900 s, 1 W at 785 nm) pumped by a green laser (Coherent Verdi V6, 532 nm, 5 W) or a tunable Ti:Sa laser system (SolsTis M Squared lasers, 1 W at 785 nm).

3. Heat killing

   To heat kill mycobacteria prior to the transfer of material from level 3 containment to a lower level for analysis, 1 mL of cell culture (up to 10^8 cfu.mL^{-1}) is placed in a heat block for 20 min at 80 °C. This protocol was validated as follows:

   After heat killing, *M. tuberculosis* was plated onto 7H10 (Sigma-Aldrich, UK) supplemented with 0.05% glycerol (Sigma-Aldrich, UK) and 10% OADC (Sigma-Aldrich, UK) and no growth was observed after 4 weeks. Completion of this protocol allows transfer of fixed *M. tuberculosis* out of the containment level 3 (CL3) laboratory to be analyzed at a lower containment level.

4. Preparation storage

   Store tissue sections at −80 °C before use. Once sliced and mounted onto quartz slides store the frozen tissue section preparation at −20 °C between two experimental days.

5. Laser power adjustment

   Adjust the power on the sample plane to obtain an optimal signal-to-noise ratio. The power used will depend on the sample being targeted. In the case of single mycobacterial cells, you can use 150 mW on the sample plane. This condition does not produce any damage to the cells during the acquisition of the spectra. However, the integration time and laser power can be adjusted depending on the cell size, specie or if a group of bacteria is targeted instead of a single cell. It is important to optimize these parameters on the different organisms in advance.

6. Standard Raman acquisition

   The acquisition time set to obtain a standard Raman spectrum is 30 s, 6 s five times accumulated. The laser wavelength is constant through the measure and set at 784.6 nm. To produce a final spectrum, a background Raman spectrum is first acquired next to a single bacterium. Once the background is taken the signal from the single cell is recorded, and the background signal is subtracted afterward. It takes 30 s to have the background signal and 30 s of acquisition to obtain the bacterial spectrum. So, in this case, it takes around 1 min in total per single bacterium standard Raman spectrum.

7. WMR spectroscopy acquisition

   During acquiring WMR spectra, each spectrum was taken at an integration time of 30 s that accumulates 6 s for five times. Five spectra were acquired continuously over 150 s when the laser was tuning over a range of 1 nm around 785 nm. So, in this case, it takes around two and a half minutes in total per single bacterium to acquire a WMR spectrum.

8. Laser wavelength calibration

   To calculate the laser wavelength using the known position of the main polystyrene peak ($\Delta\omega = 1001.4$ cm^{-1}) use the following equation:

$$\Delta\omega\left(cm^{-1}\right) = \left(\frac{1}{\lambda_0\left(nm\right)} - \frac{1}{\lambda_1\left(nm\right)}\right) \times \frac{\left(10^7\,nm\right)}{\left(cm\right)}$$

   where $\Delta\omega$ corresponds to the Raman shift, λ_0 (nm) is the laser wavelength to calculate and λ_1 (nm) the measured wavelength of the main polystyrene peak in the Raman spectrum.

9. Raman spectra analysis

   We observed that both standard and WMR spectroscopy could discriminate LR and LP cells. However, WMR spectroscopy brings both higher sensitivity and specificity [13]. The autofluorescence background is completely removed by the use of the modulation method showing the spectrum with only the Raman peaks and with the accurate ratio between peaks [14, 15]. LR mycobacteria display higher intensity in the Raman bands associated with lipids mainly at 1300 cm^{-1} and around 1440–1450 cm^{-1} [16].

10. Raman spectroscopy a label-free and non-destructive methodology

    Raman spectroscopy is label-free and non-destructive; therefore the tissue sample investigated can be used again to perform another test such as immunostaining [16]. If the tissue is studied with a staining method, the sample can be damaged meaning that it cannot be reused. However, Raman spectroscopy is an interesting method to study single bacteria both in vitro and in tissue especially if the sample is to be analyzed further using another method.

11. PCA LOOCV and Matlab

Additional information on PCA can be found in this article [17]. We used Matlab to run the PCA and the LOOCV analysis (MathWorks, UK, version R2014b).

Acknowledgment

This work was supported by the PreDiCT-TB consortium [IMI Joint undertaking grant agreement number 115337, resources of which are composed of financial contribution from the European Union's Seventh Framework Programme (FP7/2007-2013) and EFPIA companies' in kind contribution (www.imi.europa.eu). This work was supported by the Department of Health, UK. The views expressed in this chapter are those of the authors and not necessarily those of the Department of Health. This work was supported by the UK Engineering and Physical Sciences Research Council (Grant code EP/J01771X/1) and a European Union FAMOS project (FP7 ICT, 317744). Authors acknowledge the loan of a laser from M Squared Lasers.

References

1. Gillespie SH, Crook AM, McHugh TD et al (2014) Four-month moxifloxacin-based regimens for drug-sensitive tuberculosis. N Engl J Med 371(17):1577–1587

2. Merle CS, Fielding K, Sow OB et al (2014) A four-month gatifloxacin-containing regimen for treating tuberculosis. N Engl J Med 371(17):1588–1598

3. Jindani A, Harrison TS, Nunn AJ et al (2014) High-dose rifapentine with moxifloxacin for pulmonary tuberculosis. N Engl J Med 371(17):1599–1608

4. Phillips PP, Mendel CM, Burger DA et al (2016) Limited role of culture conversion for decision-making in individual patient care and for advancing novel regimens to confirmatory clinical trials. BMC Med 14:19

5. Daniel J, Deb C, Dubey VS et al (2004) Induction of a novel class of diacylglycerol acyltransferases and triacylglycerol accumulation in *Mycobacterium tuberculosis* as it goes into a dormancy-like state in culture. J Bacteriol 186(15):5017–5030

6. Garton NJ, Waddell SJ, Sherratt AL et al (2008) Cytological and transcript analyses reveal fat and lazy persister-like bacilli in tuberculous sputum. PLoS Med 5(4):e75

7. Deb C, Lee CM, Dubey VS et al (2009) A novel in vitro multiple-stress dormancy model for *Mycobacterium tuberculosis* generates a lipid-loaded, drug-tolerant, dormant pathogen. PLoS One 4(6):e6077

8. Baek SH, Li AH, Sassetti CM (2011) Metabolic regulation of mycobacterial growth and antibiotic sensitivity. PLoS Biol 9(5):e1001065

9. Hammond RJ, Baron VO, Oravcova K et al (2015) Phenotypic resistance in mycobacteria: is it because I am old or fat that I resist you? J Antimicrob Chemother 70(10):2823–2827

10. Maquelin K, Kirschner C, Choo-Smith LP et al (2002) Identification of medically relevant microorganisms by vibrational spectroscopy. J Microbiol Methods 51(3):255–271

11. Buijtels PCAM, Willemse-Erix HFM, Petit PLC et al (2008) Rapid identification of mycobacteria by Raman spectroscopy. J Clin Microbiol 46(3):961–965

12. Pahlow S, Meisel S, Cialla-May D et al (2015) Isolation and identification of bacteria by means of Raman spectroscopy. Adv Drug Deliv Rev 89:105–120

13. Canetta E, Mazilu M, De Luca AC et al (2011) Modulated Raman spectroscopy for enhanced identification of bladder tumor cells in urine samples. J Biomed Opt 16(3):037002

14. De Luca AC, Mazilu M, Riches A et al (2010) Online fluorescence suppression in modulated Raman spectroscopy. Anal Chem 82(2):738–745

15. Mazilu M, De Luca AC, Riches A et al (2010) Optimal algorithm for fluorescence suppres-

sion of modulated Raman spectroscopy. Opt Express 18(11):11382–11395

16. Baron VO, Chen M, Clark SO et al (2017) Label-free optical vibrational spectroscopy to detect the metabolic state of M. tuberculosis cells at the site of disease. Sci Rep 7(1):9844

17. Ringnér M (2008) What is principal component analysis? Nat Biotechnol 26 (3):303–304

Chapter 5

A Flow Cytometry Method for Assessing *M. tuberculosis* Responses to Antibiotics

Charlotte L. Hendon-Dunn, Stephen R. Thomas, Stephen C. Taylor, and Joanna Bacon

Abstract

Traditional drug susceptibility methods can take several days or weeks of incubation between drug exposure and confirmation of sensitivity or resistance. In addition, these methods do not capture information about viable organisms that are not immediately culturable after drug exposure. Here, we present a rapid fluorescence detection method for assessing the susceptibility of *M. tuberculosis* to antibiotics. Fluorescent markers Calcein violet-AM and SYTOX-green are used for measuring cell viability or cell death and to capture information about the susceptibility of the whole population and not just those bacteria that can grow in media postexposure. Postexposure to the antibiotic, the method gives a rapid readout of drug susceptibility, as well as insights into the concentration and time-dependent drug activity following antibiotic exposure.

Key words *Mycobacterium tuberculosis*, Antibiotic susceptibility, Calcein violet-AM, SYTOX-green, Flow cytometry

1 Introduction

Current methods for assessing the antibiotic susceptibility of *Mycobacterium tuberculosis* are lengthy and do not capture information about viable organisms that are not immediately culturable under standard in vitro conditions; as a result of antibiotic exposure [1]. We have developed a rapid dual-fluorescence flow cytometry method using markers for cell viability and death. The fluorescent markers we use are Calcein violet-AM (CV-AM; ex/em 400/452 nm) and SYTOX-green (SG; (ex/em 488 nm/523 nm). CV-AM is a dye used for distinguishing live cells through the action of intracellular esterase activity, which converts the virtually non-fluorescent cell-permeant CV-AM to the intensely fluorescent membrane impermeable Calcein violet (CV), which can be easily excited with the violet laser (and detected in channel FL6), allowing other laser lines to be used for conventional fluorochromes. SG, used for measuring cell death, permeates through damaged bacteria

Stephen H. Gillespie (ed.), *Antibiotic Resistance Protocols*, Methods in Molecular Biology, vol. 1736, https://doi.org/10.1007/978-1-4939-7638-6_5, © Springer Science+Business Media, LLC 2018

and binds to DNA. It will not cross intact membranes, but will easily penetrate compromised membranes that are characteristic of dead cells. It exhibits more than a 500-fold fluorescence enhancement upon binding nucleic acids after being excited with the blue laser (detected in channel FL1). These dyes have been used to dual-stain *M. tuberculosis* that has been exposed to a range of antibiotics with different modes of action at different concentrations over time [2]. Unlike colony counts that only capture information about bacteria that can be cultured on solid media, the flow cytometry analyses potentially capture information about non-growing populations. Recently, a combination of CV-AM and SYTOX-red was compared with other dye pairs for their ability to visualize and quantify live/dead populations during the first phase of bioadhesion in the formation of oral bacterial biofilms in a study by Tawakoli et al. [3]. More traditional plating methods were unable to quantify viable but non-culturable oral bacteria; using current approaches, over 50% of oral bacterial microbiome was un-culturable and CV-AM allowed for a clear distinction between the different susceptibility phenotypes within the biofilms [4, 5]. Another recent study has also shown that CV-AM combined with a microfluidic approach is a useful tool for gaining insights into the metabolic activity of growing and non-growing *Corynebacterium glutamicum* [6]. The flow cytometry approach has an additional advantage in that it provides insight into the mode of action of the drug; antibiotics targeting the cell wall give a distinctive flow cytometry profile compared to those inhibiting intracellular processes.

This rapid drug susceptibility method could identify more effective antimycobacterial agents, provide information about their mode of action, and aid the acceleration through the drug development pathway into the clinic.

2   Materials

2.1   Reagents for Staining

1. CV-AM (Invitrogen, Life Technologies). Gently centrifuge each tube of lyophilized CV-AM, received from the manufacturer to pellet the dye. Dissolve the pellet in 25 μL of DMSO. Minimize exposure to UV light by wrapping the tube in foil. Use freshly dissolved CV-AM on each occasion (*see* **Note 1**).

2. DMSO (Sigma Aldrich) (*see* **Note 1**).

3. SG (Invitrogen, Life Technologies) (*see* **Note 1**) Dilute each tube of SG stock solution received from the manufacturer from the concentration of 5 mM to a working solution of 20 μM in DMSO. Aliquot this and store in opaque vials at −20 °C for up to 6 months. Each vial used cannot be refrozen.

4. Hanks Balanced Salt Solution buffer (HBSS; ThermoFisher Scientific).

5. Formaldehyde (Scientific Laboratory Supplies).

6. Cyan ADP flow cytometer (Beckman Coulter) (*see* **Note 2**).

7. Sigmaplot graphical and data analysis software version 13.0 (Systat Inc.) (*see* **Note 3**).

3 Methods

3.1 Drug Susceptibility Assessment

1. Prepare individual batch cultures by inoculating 5 mL of CMM Mod2 medium [7] with cells from a mid-exponential culture of *Mycobacterium tuberculosis* to achieve an $OD_{540 \, nm}$ of 0.05.

2. To each culture, add antibiotic to achieve a series of concentrations that range from sub-inhibitory levels to several multiples of the expected minimum inhibitory concentration (MIC). The range of concentrations will be dependent on the antibiotic used. Include a control culture that contains no antibiotic. For isoniazid, the published method has used 0 μgmL^{-1}, 0.25 μgmL^{-1}, 0.5 μgmL^{-1} (MIC), 1 μgmL^{-1}, 2 μgmL^{-1}, 4 μgmL^{-1}, 8 μgmL^{-1}, 16 μgmL^{-1}, and 32 μgmL^{-1}.

3. Incubate the batch cultures at 37 °C with shaking at 200 rpm. Every 24 h, sample 450 μL of culture for CV-AM/SG staining.

4. Perform counts of colony forming units alongside the staining, to assess whether cells could be cultured on solid medium, by performing serial decimal dilutions and plating onto Middlebrook 7H10 agar + OADC [8].

3.2 Staining

1. Adjust the cell sample to an OD_{540nm} of 0.05 by diluting the cells in the growth medium that has been used in culture.

2. Stain 100 μL of bacteria with 0.5 μL CV-AM and 1 μL SG (20 μM) in each well of a 96-well microtiter plate and incubate at 37 °C for 1 h in the dark (*see* **Notes 4** and **5**).

3. After staining, spin the bacteria by centrifugation at 2885 × *g* for 2 min and resuspend in 100 μL HBSS containing a final concentration of 4% formaldehyde (v/v) (*see* **Note 6**).

3.3 Flow Cytometry

1. Examine bacteria using a flow cytometer that possesses lasers with excitatory wavelengths of 488 nm and 405 nm (*see* **Note 2**).

2. Detect SG fluorescence emission (ex/em 488 nm/523 nm) in channel FL1 [530/40 band pass (BP)], and CV-AM fluorescence (ex/em 400/452 nm) in channel FL6 (450/50 BP).

3. Analyse un-stained control samples to set a population gate around the bacteria to be analysed by using the forward scatter versus side scatter parameters.

4. Adjust voltages in channels FL1 (SG) and FL6 (CV-AM) so that the fluorescence histogram of the un-stained bacteria

appeared within the first order of the logarithmic scale of fluorescence.

5. Analyse the remaining stained samples at the settings defined in **step 4**.

6. Collect 10,000 events at a set standard 'Low' event rate.

3.4 Data Analysis

1. Analyse the acquired data to create one-parameter fluorescence histogram overlays and two-parameter dot plots (*see* **Note 7**) [8].

2. For all fluorescence dot plots, gate around the un-stained control in relation to the CV-AM fluorescence and gate around the stained live/zero drug control in relation to the SG fluorescence (*see* **Notes** 7 and **8**).

3. Obtain percentages of the total cell population residing in each gate for each parameter or each time-point (*see* **Notes** 1 and 7–9).

4. The data obtained can then be plotted as a percentage of the total population over the time-course of the experiment. Two types of graphs can be plotted either comparing different antibiotic concentrations for the same population gate or comparing different population gates for the same antibiotic concentration (*see* **Note 9**).

5. Differences between antibiotic treatments can be analysed using two-way ANOVA with appropriate post-hoc tests.

6. Inter-experiment variability can be assessed by performing a coefficient of variance test on the percentage values of the total population in each gate across three independent experiments.

4 Notes

1. The authors use Calcein violet-AM (CV-AM), cat no. C34858, and SYTOX-green (SG), Cat no. S7020 (Invitrogen, Life Technologies), which have been found to be the optimal reagents for success in this method. DMSO is used from individual 5 mL vials Cat no. D2650-5x5ML. A fresh vial should be opened each time the assay is performed.

2. The published method [2] used a CyAn ADP (9 colour) Analyser (Beckman Coulter) with attached Cytek plate loader as the capabilities of this flow cytometer possess the required specifications for these experiments, in particular 405 nm and 488 nm lasers, good resolution for detecting bacteria, and 530/40 & 450/50 BP filters.

3. The published method [2] used software package, Sigmaplot, for graphical and statistical analyses. However, other software packages, e.g., Graphpad Prism (version 6), may be used according to individual preference.

4. Larger volumes of cells can be stained by scaling up the quantity of dye used. For example, 1 mL of cells can be stained by the addition of 5 μL of the CV-AM stock solution and 10 μL of SG stock solution (20 μM). For this volume of cells, optimal staining is achieved through incubation at 37 °C for 1 h, whilst shaking at 220 rpm. The cells are left overnight to fix them prior to flow cytometry.

5. An un-stained cell sample is treated similarly to all stained samples to provide a control.

6. A fixation time of 30 min has been found to be sufficient for the sterilisation of *M. tuberculosis* (at an $OD_{540\ nm}$ of 0.5) to allow for removal from biosafety containment level 3 for flow cytometry analyses, i.e., all organisms are expected to be non-viable. Scientists should validate their own fixation step, locally, with the medium and procedures that they use.

7. The published method [2] uses Summit software version 4.3 for analyses.

8. A description of each population gate can be found in Fig. 1.

9. A worked example of the analysis process from gating strategy to graphical representation of proportion of cells within each gate can be found in Fig. 2.

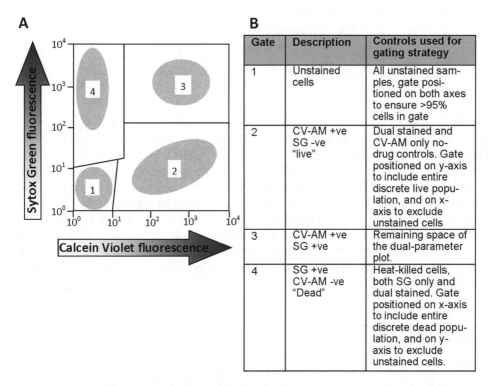

Fig. 1 (**a**) Cartoon depicting the position of the four population gates on a two parameter dot-plot with each event representing a cell in terms of its fluorescence in channel FL1 (SG) and FL6 (CV-AM). (**b**) Table describing the population gates and the controls used for positioning the gates

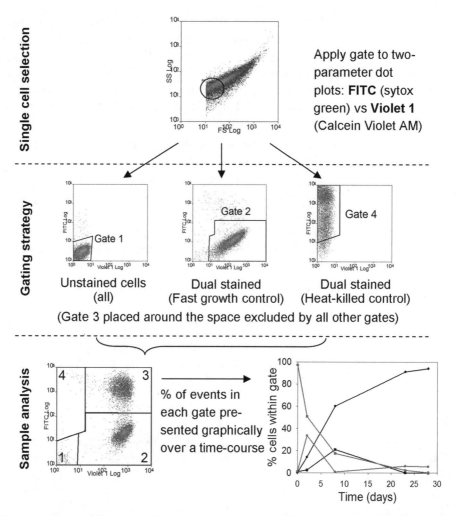

Fig. 2 Worked example of flow cytometry data capture and analysis. A single cell population is selected by placing an elliptical gate on the forward scatter vs. side scatter dot plot. This gate is applied to a FL1 vs FL6 dot-plot. Controls to aid with the subsequent gating strategy include heat-killed bacteria that are either unstained, stained with SG, or dual stained with CV-AM and SG. Polygonal population gates are placed according to the position of the populations of cells from the various controls. These controls enable the placement of gate 4, and verification that there is no spill-over of fluorescence from SG staining into the FL6 channel, which was used to measure CV-AM fluorescence intensity. Controls containing zero levels of antibiotic are stained with CV-AM only, as well as dual-stained; this enables the placement of population gate 2 and verified that there is no spill-over of fluorescence from CV-AM staining into the FL1 channel, which is used to measure SG fluorescence intensity. Percentage proportions of the total cell population within each gate are obtained and can be presented graphically

Acknowledgments

Funding was received from Department of Health Grant in Aid and the National Institute of Health Research. The views expressed in this publication are those of the authors and not necessarily those of Public Health England, the National Institute for Health Research, or the Department of Health. The research leading to this method also received funding from the Innovative Medicines Initiative Joint Undertaking under grant agreement n°115337, resources of which are composed of financial contribution from the European Union's Seventh Framework Programme (FP7/2007-2013) and EFPIA companies' in-kind contribution. We acknowledge Professor Philip Marsh for his guidance and constructive input.

References

1. Manina G, McKinney JD (2013) A single-cell perspective on non-growing but metabolically active (NGMA) bacteria. Curr Top Microbiol Immunol 374:135–161. https://doi.org/10.1007/82_2013_333

2. Hendon-Dunn CL, Doris KS, Thomas SR et al (2016) A flow cytometry method for rapidly assessing M. tuberculosis responses to antibiotics with different modes of action. Antimicrob Agents Chemother 60:3869–3883. https://doi.org/10.1128/AAC.02712-15

3. Tawakoli PN, Al-Ahmad A, Hoth-Hannig W et al (2013) Comparison of different live/dead stainings for detection and quantification of adherent microorganisms in the initial oral biofilm. Clin Oral Investig 17:841–850. https://doi.org/10.1007/s00784-012-0792-3

4. Wade WG (2013) The oral microbiome in health and disease. Pharmacol Res 69:137–143. https://doi.org/10.1016/j.phrs.2012.11.006

5. Vartoukian SR, Palmer RM, Wade WG (2010) Strategies for culture of "unculturable" bacteria. FEMS Microbiol Lett 309:1–7. https://doi.org/10.1111/j.1574-6968.2010.02000.x

6. Krämer CEM, Singh A, Helfrich S et al (2015) Non-invasive microbial metabolic activity sensing at single cell level by perfusion of calcein acetoxymethyl ester. PLoS One 10:1–24. https://doi.org/10.1371/journal.pone.0141768

7. James BW, Williams A, Marsh PD (2000) The physiology and pathogenicity of Mycobacterium tuberculosis grown under controlled conditions in a defined medium. J Appl Microbiol 88:669–677

8. Middlebrook G, Cohn ML (1958) Bacteriology of tuberculosis: laboratory methods. Am J Public Health 48:844–853. https://doi.org/10.2105/AJPH.48.7.844

Chapter 6

Application of Continuous Culture for Assessing Antibiotic Activity Against *Mycobacterium tuberculosis*

Charlotte L. Hendon-Dunn, Saba Anwar, Christopher Burton, and Joanna Bacon

Abstract

There is a proportion of the *M. tuberculosis* population that is refractory to the bactericidal action of anti-tuberculosis antibiotics due to phenotypic tolerance. This tolerance can be impacted by environmental stimuli and the subsequent physiological state of the organism. It may be the result of preexisting populations of slow growing/non replicating bacteria that are protected from antibiotic action. It still remains unclear how the slow growth of *M. tuberculosis* contributes to antibiotic resistance and antibiotic tolerance. Here, we present a method for assessing the activity of antibiotics against *M. tuberculosis* using continuous culture, which is the only system that can be used to control bacterial growth rate and study the impact of slow or fast growth on the organism's response to antibiotic exposure.

Key words Mycobacteria, Continuous culture, Chemostat, Antibiotic resistance

1 Introduction

An important aim for improving TB treatment is to shorten the period of antibiotic therapy without increasing relapse rates or encouraging the development of antibiotic-resistant strains. In any *M. tuberculosis* population there is a proportion of bacteria that are antibiotic-tolerant; this might be because of preexisting populations of slow growing/nonreplicating bacteria that are protected from antibiotic action due to the expression of a phenotype that limits antibiotic activity [1]. The methods presented here describe the use of continuous culture for assessing the activity of antibiotics against *Mycobacterium tuberculosis*, with a particular focus on the effect of either slow growth rates [69.3 h mean generation time (MGT)] or fast growth rates (23.1 h MGT) on the response of the organism to antibiotic exposure. Continuous culture is an ideal system for growing mycobacteria under defined and controlled conditions [2–5] and is the only growth system in which bacterial growth rate can be controlled. During steady-state growth in

Stephen H. Gillespie (ed.), *Antibiotic Resistance Protocols*, Methods in Molecular Biology, vol. 1736, https://doi.org/10.1007/978-1-4939-7638-6_6, © Springer Science+Business Media, LLC 2018

continuous culture the bacteria are in equilibrium with their environment and growing at a constant generation time. While the organisms are in this state, individual growth parameters can be varied independently so that the direct effect of a single stimulus (such as antibiotic action) on the viability and molecular genetics of an organism can be investigated. Antibiotic activity in *M. tuberculosis* can be assessed under a range of different environmental stimuli in continuous culture. Details of these growth conditions, which include low pH, low oxygen, and slow growth rates, have been described previously [2, 3, 6–8].

Here, bacilli replicating at the different growth rates are assessed for their responses to static concentrations of antibiotic [6, 7]. Continuous cultures can be sampled throughout a time-course (in the presence of the antibiotic) for a variety of measurements that include viability, (colony-forming units) mutation rate, changes in population genetics (whole genome sequencing), and phenotypic analyses (microarray, RNAseq, metabolomics) [2, 3, 5–9].

2   Materials

2.1 Assembly and Maintenance of the Chemostat

(*see* **Note 1**; specific equipment used by the authors)

1. 1 L Glass Vessel (DASGIP) (Eppendorf, Stevenage, UK).
2. pH probe (220 mm) plus leads (Mettler Toledo—gel filled) (Brighton Systems, New Haven, UK).
3. Buffer, reference standard, pH 4.0 ± 0.01 at 25 °C.
4. Buffer, reference standard, pH 7.0 ± 0.01 at 25 °C.
5. Broadley James Dissolved Oxygen (DO) probe (220 mm) plus leads (Brighton Systems, New Haven, UK).
6. Broadley James membrane kit (for DO probe) (Brighton Systems, New Haven, UK).
7. Broadley James filling solution (DO probe electrolyte) (Brighton Systems, New Haven, UK).
8. DOT signal amplifier (Brighton Systems, New Haven, UK).
9. Dual-wire PFA sheathed stainless steel RT probe plus leads (temperature probe) (Brighton Systems, New Haven, UK).
10. Portex silicone rubber tubing—6 × 2 mm (SLS, Nottingham, UK).
11. Sterilin silicone tubing—4 × 1.6 mm (Bore × wall) (SLS, Nottingham, UK).
12. Esco silicone tubing—1 × 2 mm (Bore × wall) (SLS, Nottingham, UK).
13. Ty-Fast cable ties—186 mm (RS components Ltd., Corby, UK).

14. Ty-Fast cable ties—141 mm (RS components Ltd., Corby, UK).

15. Tubing connectors, Y shaped for 4–5 mm tubing ID (VWR International, Lutterworth, UK).

16. Tubing connectors, Y shaped for 6–7 mm tubing ID (VWR International, Lutterworth, UK).

17. Tubing connectors, T shaped for 6–7 mm tubing ID (VWR International, Lutterworth, UK).

18. Glass media addition anti grow-back device (D.J. Lee & Co., Ferndown, UK).

19. In-line connectors (fitting, ¼″) (Cole-Parmer, Hanwell, London, UK).

20. Acro 37 TF vent device with 0.2 μm PTFE membrane (Pall Corporation, Michigan, USA).

21. Nalgene 2 L, 2125 Heavy-duty wide neck round bottles, HDPE (Jencons, Forest Row, UK).

22. 2 L glass media duran bottle (SLS, Nottingham, UK).

23. Anglicon magnetic stirrer unit (Brighton Systems, New Haven, UK).

24. Anglicon variable speed pump (Brighton Systems, New Haven, UK).

25. Watson Marlow Bredel 101U/R auto/manual control variable speed pump (0.06–2 rpm) (Watson Marlow Limited, Falmouth, UK).

26. Watson Marlow Bredel 101U/R auto/manual control variable speed pump (1.0–32 rpm) (Watson Marlow Limited, Falmouth, UK).

27. Gilson Miniplus 3 peristaltic pump (Gilson, Bedfordshire, UK).

28. 1 L glass media duran bottles (SLS, Nottingham, UK).

29. Stirring bars, PTFE, wheel—45 mm (VWR International, Lutterworth, UK).

30. Tape heater 5″ diameter (24v/50w, Brighton Systems, New Haven, UK).

31. Eycoferm fermenter controller (Eycon controller, Eurotherm-customized by Brighton Systems, New Haven, UK).

32. Eycoferm data logging software package (Brighton Systems, New Haven, UK).

33. Glass universal tubes for sampling (20 mm ID) (SLS, Nottingham, UK).

34. Titanium cabinet tubing connectors (NIS, PHE-Porton Down).

35. Digital thermometer (Tempcon Instrumentation limited, Ford, UK).

36. Clips, tubing—30 mm (VWR International, Lutterworth, UK).

37. Clips, tubing—40 mm (VWR International, Lutterworth, UK).

38. Keck ramp clamp tubing clamps—3/8″ (Cole-Parmer, Hanwell, London, UK).

39. Keck ramp clamp tubing clamps—1/4″ (Cole-Parmer, Hanwell, London, UK).

40. Bulb hand-pump (blowing ball with reservoir) (VWR International, Lutterworth, UK).

41. Nitrogen gas (BOC Medical, Worsley, UK).

42. Air pump (Brighton Systems, New Haven, UK).

43. Middlebrook 7H10 agar: The ingredients required for 1 L of Middlebrook 7H10 agar are: (15.0 g), Na_2HPO_4 (1.5 g), KH_2PO_4 (1.5 g), $(NH4)_2SO_4$ (0.5 g), L-glutamic acid (0.5 g), sodium citrate (0.4 g), ferric ammonium citrate (0.04 g), $MgSO_4 \cdot 7H_2O$ (0.025 g), $ZnSO_4 \cdot 7H_2O$ (1.0 mg), pyridoxine (1.0 mg), biotin (0.5 mg), $CaCl_2 \cdot 2H_2O$ (0.5 mg), malachite green (0.25 mg), OADC enrichment (100.0 mL), glycerol (5.0 mL). Add the glycerol to 900 mL of distilled water and add the remaining components, except for the OADC enrichment. Mix thoroughly. Gently heat and bring to the boil. Autoclave the mixture for 15 min at 121 °C. Cool to 50 °C and aseptically add 100 mL of OADC enrichment. Mix thoroughly and pour 25 mL into each petri dish. The ingredients required for 100 mL of OADC enrichment are: bovine serum albumin fraction V (5.0 g), glucose (2.0 g), NaCl (0.85 g), oleic acid (0.05 g), catalase (4.0 mg). Prepare OADC by adding all the components to the distilled water and bring the volume to 100 mL. Mix thoroughly and then filter-sterilize the solution using a 0.2 μm filter. OADC enrichment is also available as a premixed powder from BBL Microbiology Systems and Difco Laboratories.

44. CAMR Mycobacterium Medium (CMM Mod2): the ingredients required for 1 L of CMM Mod2 are ACES buffer (10.0 g), KH_2PO_4 (0.22 g), distilled water (500 mL), CMYCO solution 1 (10 mL), CMYCO solution 2 (10 mL), CMYCO solution 3 (100 mL), CMYCO solution 4 (10 mL), biotin (10 μg/mL solution) (10 mL), $NaHCO_3$ (0.042 g), glycerol (0.75 g), CMYCO solution 5 (10 mL), and Tween 80 (2 mL). Add the first two ingredients to the first volume of distilled water (500 mL). Add the remaining ingredients and solutions in the order listed. Stir the solution to dissolve all the ingredients. Adjust the pH to 6.5 with 20% KOH. Filter-sterilize the medium

using a 0.2 μm filter. Store the medium between 2 °C and 8 °C, in the dark and use within 2 months of the production date. The composition of CMYCO solution 1 per liter is $CaCl_2 \cdot 2H_2O$ (0.055 g), $MgSO_4 \cdot 7H_2O$ (21.40 g), $ZnSO_4 \cdot 7H_2O$ (2.88 g), and distilled water (1.0 L). To prepare CMYCO solution 1, add the ingredients to the water and stir to dissolve. Store it at 2–8 °C. Use it within a 6 month period from the date of production.

The composition of CMYCO solution 2 per liter is $CoCl_2 \cdot 6H_2O$ (0.048 g), $CuSO_4 \cdot 5H_2O$ (0.0025 g), $MnCl \cdot 4H_2O$ (0.002 g), concentrated HCL (0.5 mL), distilled water (1.0 L). To prepare CMYCO solution 2, add the ingredients to the water and stir to dissolve. Store it at 2–8 °C. Use it within a 6 month period from the date of production. The composition of CMYCO solution 3 per liter is L-serine (1.0 g), L-alanine (1.0 g), L-arginine (1.0 g) L-asparagine (20.0 g), L-aspartic acid (1.0 g), L-glycine (1.0 g), L-glutamic acid (1.0 g), L-isoleucine (1.0 g), L-leucine (1.0 g), distilled water (1.0 L). To prepare CMYCO solution 3, add the ingredients to the water and stir to dissolve. Store it at 2–8 °C. Use it within a week of production. The composition of CMYCO solution 4 per liter is pyruvic acid sodium salt (100.0 g) and distilled water (1.0 L). Prepare fresh CMYCO solution 4 every time CMM is prepared. Add the pyruvic acid to the water and stir to dissolve the ingredients. The composition of CMYCO solution 5 per liter is $FeSO_4 \cdot 7H_2O$ (1.0 g), concentrated HCL (0.5 mL), and distilled water (1.0 L). Prepare fresh CMYCO solution 5 every time CMM is made. Add the ingredients to the water and stir to dissolve the ingredients.

45. *Mycobacterium tuberculosis* strain *H37Rv*.

2.2 Establishing Steady-State Growth	1. Chemostat (set up as in Subheading 3.1).
	2. CMM Mod2 medium (*see* Subheading 2.1 for the recipe).
2.3 Alteration of Growth Rates in Chemostat Culture	1. Chemostat (set up as in Subheading 3.1).
	2. Spare 1 L Glass Vessel: (DASGIP) (Eppendorf, Stevenage, UK).
	3. CMM medium (*see* Subheading 2.1 for the recipe).
2.4 Addition of Antibiotics to the Culture System	1. Antibiotics.
	2. CMM Mod2 medium (*See* Subheading 2.1 for the recipe).
	3. Minisart syringe filters (0.2 μM pore; Sigma-Aldrich, Poole, UK).
	4. Sterile water.
	5. Mityvac handheld vacuum pump (mityvac.com).

2.5 Sampling and Plating M. tuberculosis to Determine Viability

1. Phosphate buffered saline (PBS) pH 7.4 (Severn Biotech Ltd., Kidderminster, UK).

2. CMM Mod2 medium (*see* Subheading 2.1 for the recipe).

3. Middlebrook 7H10 agar plates with OADC enrichment (*see* Subheading 2.1 for the recipe).

2.6 Monitoring Chemostat Parameters

1. CMM medium (*see* Subheading 2.1 described for the recipe).

2. 40% v/v Formaldehyde (SLS, Nottingham, UK).

3. Columbia blood agar plates (Biomerieux, Basingstoke, UK).

4. Tryptone soya agar plates (VWR, Lutterworth, UK).

5. Middlebrook 7H10 agar plates with OADC enrichment (*see* Subheading 2.1 for the recipe).

3   Methods

3.1 Assembling the Chemostat

Assemble the chemostat as shown in Fig. 1.

1. Fill a clean 1 L glass vessel with deionized water (approximately 500 mL) and add a magnetic stirrer bar.

2. Fit the titanium top-plate by means of the plastic threaded collar ensuring that the O-ring is in place and intact.

3. Insert oxygen probes and pH probes through the top plate by means of the titanium threaded compression ports ensuring the threaded collars are tightened to provide an airtight seal. Probes should be checked first for any damage and to ensure they are clean and that the oxygen probe contains electrolyte (*see* **Note 2**).

4. Assemble effluent and medium lines using silicon tubing and connect them to the vessel (*see* **Note 3**).

5. Assemble and connect the air inlet, the off-gas condensate bottle (Duran bottle), the acid addition lines (optional; for controlling pH), media addition lines, and sample port to the vessel. Add vent filters (0.2 μm) to air inlets, air outlets, sample port side arms, waste bottles, collection bottles and medium bottles to maintain sterility and to prevent the buildup of pressure in the vessel (*see* **Note 4**).

6. Prepare 1 M HCL in a 1 L duran bottle and autoclave it (optional; for pH control).

7. Place essential equipment in the safety cabinet, which will house the chemostat. This includes the electronic stirrer, two peristaltic pumps (medium and effluent), the CO_2 analyzer (optional), two autoclaved waste bottles, the medium bottle, the DOT amplifier, and all leads required for connection through the cabinet bulkhead.

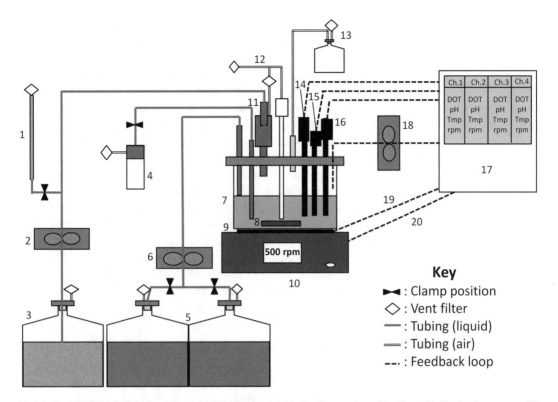

Fig. 1 Description of diagram labels: (1) Pipette assembly for flow rate calibration, (2) Media line pump, (3) Media reservoir, (4) Sample port assembly, (5) Effluent reservoirs, (6) Effluent line pump, (7) Chemostat vessel, (8) Magnetic stirrer disc, (9) Heat mat, (10) Magnetic stirrer, (11) Anti-grow-back device on media line, (12) Air Inlet assembly, (13) Air outlet/Condense bottle assembly, (14) DOT probe, (15) Temperature probe, (16) pH probe, (17) Eycoferm controller, (18) Acid addition pump controlled by feedback loop from pH probe, (19) Temperature control feedback loop, (20) DOT control feedback loop controlling stirrer speed

8. Place the vessel on top of the stirrer and on the heat pad.

9. Connect the probes to the correct channel on the Eycoferm controller, which will maintain growth parameters at set values within the culture vessel.

10. Check the oxygen probe has a signal and a rough span between 0% dissolved air saturation (DAT) (via nitrogen addition) and 100% DAT (via air addition) (*see* **Note 5**).

11. Calibrate the pH probe with pH 4.0 and pH 7.0 buffers. To control the pH (which is optional), the acid (1 M HCl) should be delivered in response to a fluctuation in acidity levels via peristaltic pumps, which will also respond to the Eycoferm controller via a negative feedback loop (*see* **Note 6**).

12. Heat the water in the vessel to 37 °C using the heat pad. To calibrate the temperature probe use a Tempcon hand-held digital thermometer to monitor the temperature in the vessel until it reaches 37 °C. Set the temperature reading on the Eycoferm controller to 37 °C (*see* **Note 7**).

13. Return all probes to the chemostat and ensure that all connections through the head plate are tightened. Pressure-test the vessel prior to autoclaving to check for air leaks. Submerge the vessel in a container of water with the probe fittings just under the water level. Vent filters should be above the water line so that they do not get wet (*see* **Note 4**). Leave all clamps in place apart from the air inlet and air outlet clamps, which should be removed. Extend the air outlet tubing past the vent filter using an additional length of tubing, with a thumb-wheel clip attached, and place the end of the tubing below the water surface. Place a 50 mL syringe onto the air inlet line and depress the plunger to push air in. Air bubbles should be seen coming out of the air outlet. Close the thumb-wheel clip on the air outlet extension and continue to push more air into the system. Observe whether air bubbles rise out of the vessel, particularly from the probe fittings on the top plate. Repair and retest any leaks that appear. Release the clamp on the air outlet and reclamp the air inlets before autoclaving.

14. Autoclave the vessel at 121 °C for 30 min to achieve sterility.

15. Ensure that waste, medium, and acid bottles are connected to the chemostat. Insert tubing into the peristaltic pumps. Connect all probes, stirrer, and CO_2 analyzer (optional) to the correct channel on the Eycoferm controller.

16. Calibrate the oxygen probe by warming up the vessel to 37 °C while stirring, and pump in nitrogen and air alternately until calibrated between 0% and 100% DAT, which is equivalent to between 0% and 20% dissolved oxygen tension (DOT) (*see* **Note 8**).

17. Switch off the heater and the stirrer. Drain the water from the chemostat.

18. Fill the vessel with 400 mL of CMM medium via the medium line. Warm the medium to 37 °C. Ensure that the stirrer unit is responding to the DAT setting on the Eycoferm controller by observing an increase in stirrer speed as the DOT level in culture decreases. Set maximum and minimum stirrer speeds.

19. Make up the inoculum by taking three confluent plates of mycobacterial colonies, which have been incubated for 2–3 weeks (*see* **Note 9**), and scrape them into 10 mL of autoclaved, distilled water. Add the inoculum through the sample port.

3.2 Establishing Steady-State Growth

1. Leave in batch mode for approximately 48 h. For the first 24 h the air inlet should be closed at the cabinet connection to allow the DAT set point of 50% DAT to be reached (*see* **Note 10**). Following this, open the air inlet as the culture will now require additional oxygen.

2. Switch the culture to a fed-batch mode by starting the medium pump at a flow rate of 5 mL/h. Set medium pump by calibrating it to the required flow rate. Keep the culture at 5 mL/h for 24 h to increase the culture volume to 500 mL (*see* **Note 11**).

3. Start the culture in continuous mode at a flow rate of 5 mL/h by switching on the effluent pump at a speed that is higher than the medium pump in order to maintain the culture volume at 500 mL. For a slow growth culture the flow rate should be maintained at this level. For a fast growth culture this flow rate is maintained for 2 days to establish the culture in continuous mode prior to an increase in flow rate (*see* **Notes 12** and **13**).

4. For a fast growth culture increase the flow rate to 10 mL/h for 2 days, and then increase the flow rate to 15 mL/h and monitor the culture daily (*see* **Notes 13** and **14**).

5. Monitor DOT levels, stirrer speed (rpm) and optical density (OD); these parameters should be stable for at least 3–5 MGT to confirm that the culture is in steady state.

6. Sample the culture for viability (cfu/mL) (*see* Subheading 3.4).

3.3 Addition of Antibiotics to the Culture System

1. Prepare antibiotics from working stocks in sterile water and pass through a 0.2 μm syringe filter.

2. Split the prepared antibiotic dilution into 2 volumes that will achieve the desired final concentration in the 500 mL chemostat and a 2 L bottle of media.

3. Aseptically add the antibiotic to a fresh 2 L bottle of media.

4. A "pre-antibiotic" culture sample should be taken at this point and processed to determine viability (*see* Subheading 3.4).

5. Drain the chemostat medium line by removing the tubing from the pump and applying a vacuum to the media bottle using a vacuum pump.

6. Connect the new medium bottle, containing the antibiotic, to the medium line and prime the line.

7. Clamp off the medium line just before the medium begins to enter the chemostat via the anti-grow-back device, reposition the tubing in the pump, and remove the clamp.

8. Add antibiotic directly to the chemostat via the sampling port, switch on the medium pump. Take a "0 h" culture sample immediately for viability analysis.

9. Check the flow rate after antibiotic addition has been commenced (*see* **Note 12**).

3.4 Sampling and Plating M. tuberculosis to Determine Viability

1. Remove a culture sample from the chemostat (*see* **Note 15**).

2. Spin 1 mL of culture at 6000 rpm (2415 x g) for 10 min.

3. Remove the supernatant and wash the pellet by resuspending it in 1 mL of PBS and spinning at 6000 rpm (2415 x g) for 10 min.

4. Remove the PBS and repeat the wash and finally resuspend in PBS

5. From the washed cell sample perform a tenfold dilution series from neat to 10^{-6} in PBS.

6. Perform the dilution series three times in parallel on each cell sample.

7. Divide agar plates into three sections (*see* **Note 16**).

8. Pipette three drops of 20 μL from each diluent onto the surface of the agar, in each plate section, using a fresh pipette tip each time.

9. Leave the plates level while the droplets dry before inverting the plates

10. Incubate the plates at 37 °C for 2–3 weeks.

11. Count the colonies.

3.5   Monitoring Chemostat Parameters

3.5.1   Daily

1. Check that the volume of liquid in the chemostat vessel is constant.

2. Check that medium is entering the vessel and effluent is going into the waste pot.

3. Check that medium and waste volumes are at the expected levels and that pumps, stirrer, and magnetic flea are all working correctly.

4. Check the waste level and swap the waste to an empty pot containing neat disinfectant if required.

5. Fill in the chemostat run sheet for temperature pH, DOT, and stirrer speed (rpm) and perform visual checks of the graphical output from the Eycoferm controller data logging.

6. Sample 5 mL of the culture for optical density. Kill the cells by the addition of 1/10 volume of 40% formaldehyde (v/v). Shake the sample vigorously and leave for 16 h before the sample can be measured for optical density. Dilute each sample fivefold in sterile, distilled water, and place the resulting cell suspension in a plastic cuvette. Read the optical density at 540 nm against water (these readings are important for determining when the culture has passed into mid-logarithmic growth and for monitoring steady-state) (*see* **Note 17**).

3.5.2   Weekly

1. Monday: Carry out culture purity checks on agar (2× blood agar and 2× Middlebrook agar plates) and measure the optical density.

2. Friday: Check the waste levels and if necessary divert the waste line to an empty waste bottle. Check that there is sufficient medium supply available for the culture to use over the weekend.

3.5.3 Occasionally

1. Move the tubing through the pumps approximately every 2 weeks in order to maintain elasticity of the tubing and to reduce the likelihood of splits developing.

2. Check flow rate (*see* **Note 12**) and temperature as required.

4   Notes

1. The authors have developed and validated the chemostat system described in this chapter and recommend the following specific items as optimal for the performance and safety of the system; 1 L Glass Vessel: DASGIP, Eppendorf, Stevenage, UK. Made to order 78903189. pH probe (220 mm) plus leads: Mettler Toledo—gel filled supplied by Brighton Systems, New Haven, UK, 104054481. Broadley James Dissolved Oxygen (DO) probe (220 mm) plus leads: Brighton Systems, New Haven, UK, N1152269. Broadley James membrane kit (for DO probe): Brighton Systems, New Haven, UK, KC1201. Broadley James filling solution (DO probe electrolyte): Brighton Systems, New Haven, UK, AS-3140-130-0025. DOT signal amplifier: Brighton Systems, New Haven, UK. Dual-wire PFA sheathed stainless steel RT probe plus leads (temperature probe): Brighton Systems, New Haven, UK, Made to order. Portex silicone rubber tubing—6 × 2 mm: SLS, Nottingham, UK, TUB3808. Sterilin silicone tubing—4 × 1.6 mm: SLS, Nottingham, UK, TUB7042. Esco silicone tubing—1 × 2 mm: SLS, Nottingham, UK, TUB7012. Tubing connectors Y shaped for 4–5 mm and 6–7 mm tubing: VWR International, Lutterworth, UK, 229-3442 and 229-3444. Tubing connectors, T shaped for 6–7 mm tubing: VWR International, Lutterworth, UK, 229-3424. Glass media addition anti grow-back device: D.J. Lee & Co., Ferndown, UK, Made to order. In-line connectors (fitting, ¼″): Cole-Parmer, Hanwell, London, UK, 06360-90. Anglicon magnetic stirrer unit: Brighton Systems, New Haven, UK, MS01. Anglicon variable speed pump: Brighton Systems, New Haven, UK. Watson Marlow Bredel 101U/R auto/manual control variable speed pump (0.06–2 rpm): Watson Marlow Limited, Falmouth, UK, 010.4002.00U. Watson Marlow Bredel 101U/R auto/manual control variable speed pump (1.0–32 rpm): Watson Marlow Limited, Falmouth, UK, 010.4202.00U. Gilson Miniplus 3 peristaltic pump: Gilson, Bedfordshire, UK, F155001, F117800. Tape heater 5″ diameter (24v/50w): Brighton Systems, New Haven, UK, Made to order. Eycoferm fermenter controller: Eycon controller, Eurotherm-customized by Brighton Systems, New Haven, UK. Eycoferm data logging software package: Brighton Systems, New Haven, UK, Made to order. Clips, tubing—30 mm: VWR

International, Lutterworth, UK, 229-0592. Clips, tubing—40 mm: VWR International, Lutterworth, UK, 229-0593. Keck ramp clamp tubing clamps—3/8″: Cole-Parmer, Hanwell, London, UK, KH-06835-07. Keck ramp clamp tubing clamps—1/4″: Cole-Parmer, Hanwell, London, UK, KH-06835-03.

2. Store the pH probes in 3 M KCl and rinse with distilled water before use. Oxygen probes are stored dry. Change the membrane on the oxygen probe and refill with fresh electrolyte when the membrane has lost integrity and/or when the electrolyte level is low.

3. There are three thicknesses of tubing. To achieve a precise flow rate the internal diameter of the tubing that runs through the medium pump head should have a bore width of no more than 1 mm. However, thicker tubing with a wider bore width of 4 mm is used for the medium and effluent lines, and the thickest tubing with a bore width of 6 mm is used for all gaseous inlets and outlets.

4. Dead ends (some of which will be used for drawing off medium or cell samples), such as air outlets, sample port side arms, waste bottles, collection bottles and medium bottles, flow rate measurement devices, also need to be fitted with vent filters to ensure that gas can be released continuously from the culture during growth and pressure does not build up in the vessel. The air outlet/off-gas line can be fed into a CO_2 analyzer for measurement of CO_2 levels.

5. Do not calibrate oxygen probes before autoclaving because the electrolyte is affected by the heat. Oxygen is only transferred evenly throughout the culture if it is stirred or shaken effectively. Standing cultures of *M. tuberculosis* will result in microenvironments in which the oxygen levels will be very low reaching microaerophilic or anaerobic levels.

6. There is a pH/temperature compensation mode on the Eycoferm controller to compensate for temperature differences because pH calibration is done at room temperature. Fittings that are exposed to the acid or alkali should be made of inert metal such as titanium to prevent corrosion caused particularly by the acid.

7. The heat pad is electronically controlled by the Eycoferm controller unit to maintain temperature at 37 °C.

8. The dissolved oxygen tension (DOT) is maintained by an immediate response of the Eycoferm controller to a drop in oxygen level, which in turn alters the stirrer speed to draw more air into the medium. The Eycoferm controller unit automatically controls the extent to which the culture is stirred via a magnetic stirring device and a flea. A DOT of 10% is equiva-

lent to 50% dissolved air saturation (DAT). The Eycoferm controller displays the DAT and not the DOT.

9. It is not advisable to use plates that are more than 3 weeks old because growth in the chemostat will be slow and cells will be more clumped in culture.

10. The DOT level in the culture could be above 10% DOT (10% DOT is equivalent to 50% DAT) and will need to drop to 10% DOT as soon as possible. High DOT levels are indicative of poor or slow growth. It is important for the stirrer speed to increase to disperse the cells. Once the DOT has dropped to the set point (10%), the Eycoferm controller will inform the stirrer to increase its speed to maintain a DOT of 10%. A consistent DOT level and a high stirrer speed are indicative of active growth.

11. DOT levels may fluctuate during a transition from batch to fed-batch.

12. A high flow rate too early on in continuous mode may lead to culture "wash-out". The flow rate is measured using a device that is constructed using a glass pipette which has been inserted into the tubing between the medium bottle and the medium pump via a connector with a T junction in it. The pipette is capped with a piece of tubing and a vent filter (Fig. 1). The bottom of the pipette is normally clamped off. The clamp is removed and medium is drawn up into the pipette using a syringe attached to the vent filter at the top of the pipette. The medium bottle is then clamped off so that the culture subsequently draws the medium from the pipette and not from the medium bottle. The speed at which the culture uses the medium from the pipette is then measured. The flow rate and dilution rates can then be calculated. Remember to remove the clamp from the medium bottle and replace the clamp at the bottom of the pipette once flow rate determinations have been completed.

13. A flow rate of 5 mL/h will give a dilution rate of 0.01 h^{-1} and a mean generation time of 69.3 h (slow growth). Whereas a flow rate of 15 mL/h will give a dilution rate of 0.03 h^{-1} and a mean generation time of 23.1 h (fast growth). The flow of medium into the vessel (F) is related to the culture volume (V) by the dilution rate (D) where $D = F/V$. The volume is expressed in mL, the flow rate is expressed in mL/h, so that the dilution rate is therefore expressed as h^{-1}. Under steady-state conditions the biomass remains constant, therefore the specific growth rate (μ) must equal the dilution rate, i.e., $\mu = D$. The dilution rate is related to the doubling time (T_d) by the equation $T_d = \log_e 2/D$. The equation, ($X = Y_{x/s} (S_o - S)$), has been derived from the material balance equation of the limit-

ing nutrient across the system. The relationship between the yield of cells using substrate, $\Upsilon_{x/s}$ (grams of biomass per gram of substrate), and the limiting substrate can be calculated using $X = \Upsilon_{x/s} (S_0 - S)$, where X is the biomass at steady state (g/L), and S_0 and S are the concentrations of the limiting substrate in the feed and residual substrate in the outflow respectively.

14. Run sheets and the Eycoferm controller data logging system are used to record parameters on a daily basis.

15. The sampling regimen that has been used in previous experiments is to sample at each mean generation time in steady-state pre and post antibiotic addition; for MGT of 69.3 h (slow growth) this is every 3 days and for a MGT of 23.1 h this is every day. The antibiotics that have been used for previous studies were isoniazid, rifampicin, and pyrazinamide. These antibiotics were used at minimum inhibitory concentrations [6, 7].

16. Plates must be preincubated at room temperature for at least several hours (although not overnight as this causes overdrying of the plates).

17. Under certain growth conditions mycobacteria will adhere to the walls of the vessels and the probes. Once this starts to occur, the optical density is likely to fall and the DOT and pH levels will fluctuate. The culture will no longer be in steady state and will need to be transferred to another vessel.

Acknowledgments

Funding was received from Department of Health Grant in Aid and the National Institute of Health Research. The views expressed in this publication are those of the authors and not necessarily those of Public Health England, the National Institute for Health Research, or the Department of Health. The research leading to these results also received funding from the Innovative Medicines Initiative Joint Undertaking under grant agreement n°115337, resources of which are composed of financial contribution from the European Union's Seventh Framework Programme (FP7/2007-2013) and EFPIA companies' in-kind contribution. The authors acknowledge Kim Hatch and Jon Allnutt for the huge contribution they have made to the development of chemostat models for the growth of *M. tuberculosis* and application to the assessment of antibiotic activity and to Professor Philip Marsh for his guidance and constructive input.

References

1. Aldridge BB, Keren I, Fortune SM (2014) The spectrum of drug susceptibility in mycobacteria. Microbiol Spectr 2:1–14. https://doi.org/10.1128/microbiolspec.MGM2-0031-2013

2. Bacon J, James BW, Wernisch L et al (2004) The influence of reduced oxygen availability on pathogenicity and gene expression in *Mycobacterium tuberculosis*. Tuberculosis (Edinb) 84:205–217. https://doi.org/10.1016/j.tube.2003.12.011

3. Bacon J, Dover LG, KA H et al (2007) Lipid composition and transcriptional response of *Mycobacterium tuberculosis* grown under iron-limitation in continuous culture: identification of a novel wax ester. Microbiology 153:1435–1444. https://doi.org/10.1099/mic.0.2006/004317-0

4. Golby P, Hatch KA, Bacon J et al (2007) Comparative transcriptomics reveals key gene expression differences between the human and bovine pathogens of the *Mycobacterium tuberculosis* complex. Microbiology 153:3323–3336. https://doi.org/10.1099/mic.0.2007/009894-0

5. Beste DJV, Espasa M, Bonde B et al (2009) The genetic requirements for fast and slow growth in mycobacteria. PLoS One 4:e5349. https://doi.org/10.1371/journal.pone.0005349

6. Jeeves RE, Marriott AA, Pullan ST et al (2015) Mycobacterium tuberculosis is resistant to isoniazid at a slow growth rate by single nucleotide polymorphisms in katG codon Ser315. PLoS One 10:e0138253. https://doi.org/10.1371/journal.pone.0138253

7. Pullan ST, Allnutt JC, Devine R et al (2016) The effect of growth rate on pyrazinamide activity in *Mycobacterium tuberculosis* – insights for early bactericidal activity? BMC Infect Dis 16:205. https://doi.org/10.1186/s12879-016-1533-z

8. Jenkins C, Bacon J, Allnutt J et al (2009) Enhanced heterogeneity of rpoB in *Mycobacterium tuberculosis* found at low pH. J Antimicrob Chemother 63:1118–1120. https://doi.org/10.1093/jac/dkp1259.

9. Beste DJ, Hooper T, Stewart G et al (2007) GSMN-TB: a web-based genome-scale network model of *Mycobacterium tuberculosis* metabolism. Genome Biol 8:R89. https://doi.org/10.1186/gb-2007-8-5-r89

Chapter 7

Real-Time Digital Bright Field Technology for Rapid Antibiotic Susceptibility Testing

Chiara Canali, Erik Spillum, Martin Valvik, Niels Agersnap, and Tom Olesen

Abstract

Optical scanning through bacterial samples and image-based analysis may provide a robust method for bacterial identification, fast estimation of growth rates and their modulation due to the presence of antimicrobial agents. Here, we describe an automated digital, time-lapse, bright field imaging system (oCelloScope, BioSense Solutions ApS, Farum, Denmark) for rapid and higher throughput antibiotic susceptibility testing (AST) of up to 96 bacteria–antibiotic combinations at a time. The imaging system consists of a digital camera, an illumination unit and a lens where the optical axis is tilted 6.25° relative to the horizontal plane of the stage. Such tilting grants more freedom of operation at both high and low concentrations of microorganisms. When considering a bacterial suspension in a microwell, the oCelloScope acquires a sequence of 6.25°-tilted images to form an image Z-stack. The stack contains the best-focus image, as well as the adjacent out-of-focus images (which contain progressively more out-of-focus bacteria, the further the distance from the best-focus position). The acquisition process is repeated over time, so that the time-lapse sequence of best-focus images is used to generate a video. The setting of the experiment, image analysis and generation of time-lapse videos can be performed through a dedicated software (UniExplorer, BioSense Solutions ApS). The acquired images can be processed for online and offline quantification of several morphological parameters, microbial growth, and inhibition over time.

Key words Automated digital time-lapse bright field screening system, oCelloScope, Qualitative and quantitative image-based analysis, Generation of time-lapse videos, UniExplorer, Bacterial cultures and clinical isolates, Antibiotic resistance testing

1 Introduction

As consequence of the dramatic increase in microbial resistance and the recurrent need for treatment with newer and often more expensive antibiotics, the ability to develop cost-effective, fast and accurate antimicrobial susceptibility testing (AST) methods is currently in the spotlight [1, 2]. The most widely used AST methods include manual tests such as disk diffusion and broth microdilution [3–5], as well as phenotypic [6–9] and genotypic [10–12] techniques. Manual tests provide flexibility, possible cost saving

Stephen H. Gillespie (ed.), *Antibiotic Resistance Protocols*, Methods in Molecular Biology, vol. 1736, https://doi.org/10.1007/978-1-4939-7638-6_7, © Springer Science+Business Media, LLC 2018

and quantitative results (e.g., determination of minimum inhibitory concentration) [1, 13], although they may not accurately predict the results of many clinical samples [14]. Additionally, new emerging mechanisms of bacterial resistance require continuous revision of the adequacy of each AST method [15, 16].

We developed an automated digital, time-lapse, bright field imaging system (oCelloScope, BioSense Solutions ApS, Farum, Denmark) for microbiological research allowing rapid AST (Fig. 1). It allows fast monitoring of up to 96 bacteria–antibiotic combinations in a time as quick as 2 min 19 s when a single image per well is acquired. The oCelloScope supports several types of samples and containers including microscope slides and microtiter plates up to 96 wells. It is well suited for liquid samples such as bacterial cultures and clinical isolates such as urine [17] and blood [18] samples. Moreover, the oCelloScope also showed promising preliminary results for solid cultures of a single bacterial colony growing on a semitransparent medium. The miniaturized imaging system comprises a digital camera, an illumination unit, and a lens where the optical axis is tilted 6.25° relative to the horizontal plane of the stage (Fig. 2). Such tilt allows scanning of volumes (Fig. 2) and formation of an image Z-stack containing the best-focus image, as well as the adjacent out-of-focus images (which contain progressively more out-of-focus bacteria, the further the distance from the best-focus position). When all the bacteria are sedimented in a microwell, they are all in focus in the best-focus image (Fig. 2). An optical resolution of 1.3 μm is achieved through a proprietary lens system and a monochrome 5 megapixel complementary metal oxide semiconductor (CMOS) camera chip (5.6 mm length and 4.2 mm height) with a focus depth of ~10 μm and an optical magnification factor of 4. The dimensions of the oCelloScope (45 × 26 × 25 cm) allow portability and operation in standard laboratory incubators for thorough control of temperature (20–40 °C) and humidity (20–93%). The oCelloScope comprises UniExplorer software (BioSense Solutions ApS) for both setting the experiment and data analysis. The software communicates with the instrument via an Ethernet connection, which can favorably reduce the time required in the laboratory when working with hazardous microorganisms. The UniExplorer software generates time-lapse videos of the best-focus images acquired over time and performs image-based analysis for online and offline quantification of several morphological parameters [18–22] and microbial growth and inhibition over time [23, 24]. The output data can also be exported to Microsoft Excel or in the CSV format for further processing.

AST using the oCelloScope showed a statistically significant antibiotic effect within 6 min for *Escherichia coli* monocultures and within 30 min in complex samples from pigs with catheter-associated urinary tract infections [19]. Additional investigations

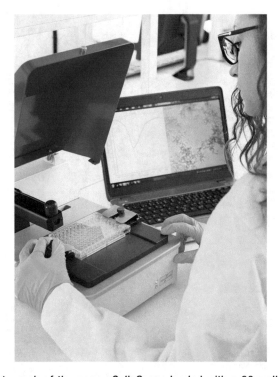

Fig. 1 Photograph of the open oCelloScope loaded with a 96-well plate. The UniExplorer software allows setting the experiment as well as performing online and offline image analysis along with generation of time-lapse videos. Growth/growth inhibition curves can be inspected and compared among the different wells

demonstrated the suitability of the system to early detect the resistance profiles of bacteria reference strains and multidrug-resistant clinical isolates [17], as well as fungal strains [20]. Furthermore, image analysis performed with the oCelloScope was shown to allow measuring bacterial length and morphological changes and, hence, differentiating between normal growth patterns and bacterial filamentation. This would otherwise be impossible using conventional optical density measurements [18]. This is particularly relevant when testing β-lactam antibiotics such as penicillins, cephalosporins, carbapenems, and monobactams, which typically induce morphological changes in bacteria such as filamentation and spheroplast formation [25–27].

2   Materials

- For bacteria isolated from pure cultures: prepare sterile media according to the strain nutritional requirements [28].

- For urine and blood samples: use cation-adjusted Müller-Hinton broth (MH broth) with *N*-tris(hydroxymethyl)methyl-

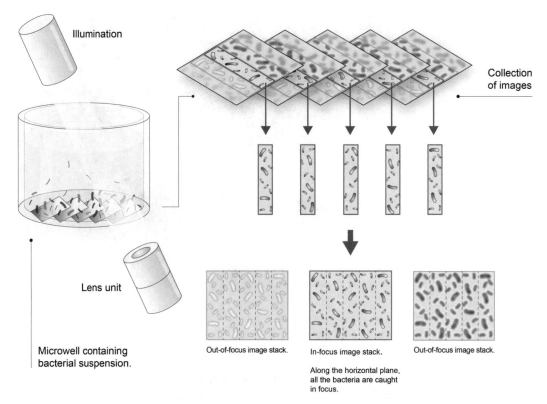

Illumination

Collection
of images

Lens unit

Microwell containing
bacterial suspension.

Out-of-focus image stack.

In-focus image stack.

Along the horizontal plane,
all the bacteria are caught
in focus.

Out-of-focus image stack.

Fig. 2 Schematic showing the oCelloScope optical scanning technology applied to a microwell containing a bacterial suspension. The optical axis of the miniaturized imaging system is tilted 6.25° relative to the horizontal plane of the stage to allow scanning of volumes. Images are acquired to form an image Z-stack. When all the bacteria are sedimented in a microwell, they are all in focus in the best-focus image

2-aminoethanesulfonic acid (TES) (CAMHBT, Sensititre®, Thermo Scientific) [17, 18].

- Transparent flat bottom microtiter plates with up to 96 wells (e.g., Corning® Costar®, Sigma-Aldrich, product No. CLS3596).

- Densitometer to determine the starting concentration of bacterial suspensions via optical density measurements at 600 nm (OD_{600}).

3   Methods

Subheading 3.1 refers to a general guideline for preparation of pure bacterial strain cultures [19], whereas Subheadings 3.2 and 3.3 refer to examples of preparation procedures for cultures from urine [17] and blood [18] samples, respectively. The suggested procedures can be modified and further optimized depending on the sample source and the bacterial strains. During sample preparation, biological waste management guidelines should be followed.

3.1 Guideline for Preparation of Pure Cultures of Bacterial Strains

1. Inoculate the bacterial strain of interest in 3–5 mL of appropriate growth medium according to its metabolic and environmental requirements.

2. Inoculate 0.1 mL of an overnight culture in 8 mL of medium for 2 h at 37 °C to reach the exponential phase.

3. Standardize bacterial suspensions by adjusting the concentration of the inocula to ca. 2×10^7 bacteria/mL using OD_{600}.

4. Dilute to a final concentration of 7.5×10^5 bacteria/mL.

5. Load the bacterial suspension in a transparent flat bottom 96-well plate.

6. Prepare the antibiotic dilution range in growth medium using series of doubling dilution (e.g., 16, 8, 4, 2, 1, 0.5 µg/mL) and add a proper volume to each well.

3.2 Guideline for Preparation of Urine Samples

1. Centrifuge urine samples at $2400 \times g$ for 5 min at room temperature to remove leukocytes and epithelial cells.

2. Collect the supernatant, mix 50% (vol/vol) with CAMHBT and incubate for 2 h at 37 °C.

3. Adjust bacterial concentration to 0.5 of the McFarland standard.

4. Inoculate 0.1 mL in 11 mL of MH broth and mix thoroughly.

5. Load the bacterial suspension in a transparent flat bottom 96 well plate. Prepare the antibiotic dilution range in growth medium using series of doubling dilution (e.g., 16, 8, 4, 2, 1, 0.5 µg/mL) and add a proper volume to each well.

3.3 Guideline for Preparation of Blood Samples

1. Centrifuge blood samples at $200 \times g$ for 5 min to remove blood cells.

2. Resuspend the bacterial pellet in CAMHBT and incubate for 2 h at 37 °C.

3. Inoculate 0.1 mL in 11 mL of MH broth and mix through.

4. Load the bacterial suspension in a transparent flat bottom 96-well plate.

   Prepare the antibiotic dilution range in growth medium using series of doubling dilution (e.g., 16, 8, 4, 2, 1, 0.5 µg/mL) and add a proper volume to each well.

3.4 Automated Time-Lapse Analysis Using the oCelloScope System

1. Place the oCelloScope in a standard laboratory incubator for biological applications for precise temperature regulation. Prior performing the AST experiment, it is recommended to let the instrument and the sample container equilibrate inside the incubator for at least 2 h. This step allows the prevention of condensation forming on the on the microtiter plate lid,

which would otherwise affect the quality of the acquired images and, hence, the quantitative analysis (*see* **Note 1**).

2. Introduce the 96 well plate loaded with the samples in the plate holder, tight firmly and close the instrument lid (*see* **Note 2–4**).

3. Configure the UniExplorer software for time-lapse analysis by clicking the "New job" button in the top left corner of the main window.

4. Give the new job a name (*see* **Note 5**).

5. Choose the proper acquisition modules from the list to the left by double clicking (or dragging) the icons in the following order: (a) "Acquire" (for recording new image data) and (b) "Growth Kinetics Analysis" (for monitoring bacterial growth/ growth inhibition over time). Press 'Next'.

6. Let the UniExplorer software recognize the oCelloScope instrument by selecting the instrument to use from the instrument list with a left click. If the instrument is not displayed on the list, check cables and power. If the instrument is still not displayed, press the button 'Refresh instrument list'. Otherwise, the IP address of the instrument (e.g., 012.34.56.789) can be manually specified by typing it in the space at the bottom of the window. Press 'Next' to continue setting the experiment setup.

7. Select the sample container type (e.g., "96 wells, Costar® Corning 3596") from the list showing the type of sample containers supported by the UniExplorer software.

8. Select and enable the wells that should be included in the analysis using the cursor. Enabled wells are shown as blue. Bacterial growth/growth inhibition is only monitored for enabled wells.

9. The UniExplorer software automatically sets focus and the optimal illumination level for each well. If necessary this can manualy be adjusted for each single well with the "Live View".

10. Set the scan area for each well. The scan area is shown as an orange rectangle and placed at the center of each well by default. Multiple scan areas can be added in different positions in each well and labeled with customized names. The number of images acquired per scan area can be specified and the image distance is set according to the objects size. An image distance of 4.9 μm should be chosen for microorganisms.

11. Adjust the time of analysis by selecting the number of acquired images ("Number of repetitions") and the time interval between two sequential images ("Repetition interval"). For instance, by selecting "Number of repetitions" = 33 and "Repetition interval" = 00:15:00, the oCelloScope will take the images every 15 min for 8 h with the first image taken at $t = 0$. By ticking "Use multiple repetition intervals," it is pos-

sible to set two different phase of analysis with different time intervals between sequential images. Such intervals can be adjusted according to the expected phases of bacterial growth, depending on the microorganism type. The acquisition time ("Acquire time") refers to the overall time required for both recording the images and analyze the data. Therefore, the acquisition time may exceed the actual time required for acquiring the images due to data processing. Once the time of analysis has been set, press 'Next'.

12. Select or deselect the algorithms to use for image analysis by ticking the boxes. Each algorithm is designed to give specific advantages depending on analysis type and sample properties, such as cell concentration and translucency.

- The Total Absorption (TA) algorithm is an equivalent of OD_{600}. During microbial growth, the increasing number of bacteria will reduce light transmission through the sample and the image will get progressively darker. Any darker image corresponds to a higher TA value. TA sensitivity is limited if compared to the BCA algorithm (described below) as growth and cell concentration need to be quite considerable before affecting light transmission.

- The Background Corrected Absorption (BCA) algorithm is an equivalent of OD_{600} but with increased sensitivity even at very low or high cell concentrations. To achieve such performance, the BCA algorithm considers any variation in the background intensity relative to the first acquired image. Hence, an even light distribution in the images can be obtained, which is used for calculating a threshold pixel value. Such threshold value divides image pixels into pixels belonging to the background and pixels belonging to the microorganisms. Growth curves are generated based on changes in the pixels belonging to the microorganisms. Therefore, the BCA algorithm is able to determine microbial growth/growth inhibition with high sensitivity as the influence of the background intensity is significantly reduced compared to the same analysis performed with the TA algorithm.

- The Segmentation and Extraction of Surface Area (SESA) algorithm identifies all the objects in a scan based on their contrast with the background and then it calculates the total surface area covered by such objects. It is insensitive to variations in the background intensity (caused by for example condensation on microtiter plate lid) and it is able to measure microbial growth with high accuracy at very low cell concentrations. However, when more than 20% of the total image area is covered with bacteria, the algorithm accuracy starts to decline.

- The Normalized version of each algorithm (Norm) is computed in the same way as the respective algorithm but the value of the first image is subtracted from the subsequent ones to build the growth/growth inhibition curve.

- Once the algorithms for image analysis have been chosen, press "Start" to start the time-lapse AST experiment.

13. Real-time growth/growth inhibition curves ("Growth Kinetics Analysis") and image data ("Acquire") can be monitored during the AST experiment in the "Current scan" tab. Notes about the experiment can be added in the "Job overview" tab, whereas the "Saved images" tab shows all the images that have already been acquired for each scan area as either images or time-lapse videos. In the same tab, curves for different scan areas can be compared by (a) selecting "All items" from the "Select scan area" drop down list and (b) ticking the scan areas of interest in the list to the right of the displayed graph. Each curve can be inspected in combination with the corresponding time-lapse video by ticking the box "Show video."

14. Export growth/growth inhibition curves and image data. Press the "Export" button and select "Export chart," "Export image," or "Export video" to export graphs as PNG files, images as BMP files and time-lapse videos in the AVI format. To export growth/growth inhibition values as Excel files or in the CSV format, press the "Export" button and select "Export to Excel" or "Export to CSV," respectively.

4   Notes

1. When analyzing samples with few and/or small bacteria, it may be beneficial to add polystyrene beads of an appropriate size (e.g., 3K/4K series particle counter standards, Thermo Scientific) to facilitate the focusing process [19]. During the analysis, the segmentation algorithm will be able to ignore the beads.

2. The full list of compatible plate formats is reported in the oCelloScope user manual. The oCelloScope is also compatible with microscope slides using the specific slide holder.

3. When preparing multiple samples in a microtiter plate, it is recommended to add the same volume to each well. This allows the same settings for illumination and focus to be applied to all wells.

4. The recommended total volume of bacterial suspension and antibiotic solution to use is 50–200 μL. Smaller volumes can

be used as long as the sample fully covers the bottom of the well during the entire period of analysis.

5. It is always recommended to run the experiment in triplicate including positive controls, where the same volume of plain growth medium is added. Positive control wells are expected to show bacterial growth over time.

References

1. Jorgensen JH, Ferraro MJ (2009) Antimicrobial susceptibility testing: a review of general principles and contemporary practices. Clin Infect Dis 49:1749–1755

2. Jenkins SG, Schuetz AN (2012) Current concepts in laboratory testing to guide antimicrobial therapy. Mayo Clin Proc 87:290–308

3. Ge B, Wang F, Sjölund-Karlsson M et al (2013) Antimicrobial resistance in Campylobacter: susceptibility testing methods and resistance trends. J Microbiol Methods 95:57–67

4. Berghaus LJ, Giguère S, Guldbech K et al (2015) Comparison of Etest, disk diffusion, and broth macrodilution for in vitro susceptibility testing of Rhodococcus equi. J Clin Microbiol 53:314–318

5. Baker CN, Stocker SA, Culver DH et al (1991) Comparison of the E test to agar dilution, broth microdilution, and agar diffusion susceptibility testing techniques by using a special challenge set of bacteria. J Clin Microbiol 29:533–538

6. Dortet L, Poirel L, Nordmann P (2015) Rapid detection of ESBL-producing enterobacteriaceae in blood cultures. Emerg Infect Dis 21:504–507

7. Van Belkum A, Dunne WM (2013) Next-generation antimicrobial susceptibility testing. J Clin Microbiol 51:2018–2024

8. Ahmed MAS, Bansal D, Acharya A et al (2016) Antimicrobial susceptibility and molecular epidemiology of extended-spectrum beta-lactamase-producing Enterobacteriaceae from intensive care units at Hamad Medical Corporation, Qatar. Antimicrob Resist Infect Control 11:1–6

9. Mohan R, Mukherjee A, Sevgen SE et al (2013) A multiplexed microfluidic platform for rapid antibiotic susceptibility testing. Biosens Bioelectron 49:118–125

10. Liu T, Lu Y, Gau V et al (2014) Rapid antimicrobial susceptibility testing with electrokinetics enhanced biosensors for diagnosis of acute bacterial infections. Ann Biomed Eng 42:2314–2321

11. Celandroni F, Salvetti S, Gueye SA et al (2016) Identification and pathogenic potential of clinical bacillus and paenibacillus isolates. PLoS One 11:0152831

12. Waldeisen JR, Wang T, Mitra D et al (2011) A real-time PCR antibiogram for drug-resistant sepsis. PLoS One 6:e28528

13. Wiegand I, Hilpert K, Hancock REW (2008) Agar and broth dilution methods to determine the minimal inhibitory concentration (MIC) of antimicrobial substances. Nat Protoc 3:163–175

14. Doern GV (2011) Antimicrobial susceptibility testing. J Clin Microbiol 49:S4

15. Turnidge J, Paterson DL (2007) Setting and revising antibacterial susceptibility breakpoints. Clin Microbiol Rev 20:391–408

16. Depalma G, Turnidge J, Craig BA (2016) Determination of disk diffusion susceptibility testing interpretive criteria using model-based analysis: development and implementation. Diagn Microbiol Infect Dis. https://doi.org/10.1016/j.diagmicrobio.2016.03.004

17. Fredborg M, Rosenvinge FS, Spillum E et al (2015) Rapid antimicrobial susceptibility testing of clinical isolates by digital time-lapse microscopy. Eur J Clin Microbiol Infect Dis 34:2385–2394

18. Fredborg M, Rosenvinge FS, Spillum E et al (2015) Automated image analysis for quantification of filamentous bacteria. BMC Microbiol 15:1–8

19. Fredborg M, Andersen KR, Jorgensen E et al (2013) Real-time optical antimicrobial susceptibility testing. J Clin Microbiol 51:2047–2053

20. Aunsbjerg SD, Andersen KR, Knøchel S (2015) Real-time monitoring of fungal inhibition and morphological changes. J Microbiol Methods 119:196–202

21. Kjeldsen T, Sommer M, Olsen JE (2015) Extended spectrum β-lactamase-producing Escherichia coli forms filaments as an initial response to cefotaxime treatment. BMC Microbiol 15:1–6

22. Jelsbak L, Mortensen MIB, Kilstrup M et al (2016) The in vitro redundant enzymes PurN and PurT are both essential for systemic infection of mice in Salmonella enterica serovar Typhimurium. Infect Immun 84:2076–2085

23. Khan DD, Lagerbäck P, Cao S et al (2015) A mechanism-based pharmacokinetic/pharmacodynamic model allows prediction of antibiotic killing from MIC values for WT and mutants. J Antimicrob Chemother 70:3051–3060

24. Uggerhøj LE, Poulsen TJ, Munk JK et al (2015) Rational design of alpha-helical antimicrobial peptides: do's and don'ts. Chembiochem 16:242–253

25. Yao Z, Kahne D, Kishony R (2012) Distinct single-cell morphological dynamics under beta-lactam antibiotics. Mol Cell 48:705–712

26. Periti P, Nicoletti P (1999) Classification of betalactam antibiotics according to their pharmacodynamics. J Chemother 11:323–330

27. Greenwood D, O'Grady F (1973) Comparison of the responses of Escherichia coli and Proteus mirabilis to seven β-lactam antibiotics. J Infect Dis 128:211–222

28. Goldman E, Green LH (2015) Practical handbook of microbiology. CRC Press, Boca Raton, FL

Chapter 8

Enhanced Methodologies for Detecting Phenotypic Resistance in Mycobacteria

Robert J.H. Hammond, Vincent O. Baron, Sam Lipworth, and Stephen H. Gillespie

Abstract

Lipid droplets found in algae and other microscopic organisms have become of interest to many researchers partially because they carry the capacity to produce bio-oil for the mass market. They are of importance in biology and clinical practice because their presence can be a phenotypic marker of an altered metabolism, including reversible resistance to antibiotics, prompting intense research.

A useful stain for detecting lipid bodies in the lab is Nile red. It is a dye that exhibits solvatochromism; its absorption band varies in spectral position, shape and intensity with the nature of its solvent environment, it will fluoresce intensely red in polar environment and blue shift with the changing polarity of its solvent. This makes it ideal for the detection of lipid bodies within *Mycobacterium* spp. This is because mycobacterial lipid bodies' primary constituents are nonpolar lipids such as triacylglycerols but bacterial cell membranes are primarily polar lipid species. In this chapter we describe an optimal method for using Nile red to distinguish lipid containing (Lipid rich or LR cells) from those without lipid bodies (Lipid Poor or LP). As part of the process we have optimized a method for separating LP and LR cells that does not require the use of an ultracentrifuge or complex separation media. We believe that these methods will facilitate further research in these enigmatic, transient and important cell states.

Key words Tuberculosis, Dormancy, Phenotypic resistance, Lipid body

1 Introduction

In recent years there has been an increasing interest in mycobacteria within which lipid bodies are seen [1–3]. This is due to the important association with low metabolic state and phenotypic resistance to key anti-tuberculosis antibiotics [4–7]. As the goal of improving tuberculosis treatment remains frustratingly out of reach, it is important that we understand what the true susceptibility of *M. tuberculosis* is as it is clear there is a significant difference in the susceptibility of cells with lipid bodies present in comparison with those that are not [8]. It is of considerable importance, therefore, to be able to reliably separate and quantify mycobacterial cells in different cells state. Previously published methods are effective

Stephen H. Gillespie (ed.), *Antibiotic Resistance Protocols*, Methods in Molecular Biology, vol. 1736,
https://doi.org/10.1007/978-1-4939-7638-6_8, © Springer Science+Business Media, LLC 2018

but often complex and may result in metabolic alteration in the cells studied.

Separation of particles based on their buoyant density has practiced since at least 1919 [9]. Differences in buoyant density can be used to separate particles [10–13], and the density-dependent cell sorting (DDCS) method has been applied to laboratory cultured bacteria [14]. Cells in different physiological states have been successfully separated using this approach [15] because physiological changes alter cellular components and the subsequent buoyant density. The DDCS method has been applied mostly to pure cultures [16]. It can be used as a purifying process for differential centrifugation. For mycobacteria, methods to permit separation of LR and LP cells have been described [17, 18]. Equilibrium sedimentation classically uses a gradient of a solution such as sucrose to separate particles based on their individual densities. These usually require extended centrifugation, ultracentrifugation or the use of complex separation media such as sucrose or Percoll [19]. Very little is known about the effect of these processes on the metabolism of mycobacteria, which is often the purpose of the experiments. Sucrose separation gradients can provide a carbon source for mycobacteria that are not fastidious and can utilize almost all simple carbohydrate carbon sources including sucrose [20].

Isopycnic centrifugation refers to a method wherein a density gradient is either preformed or forms during high speed centrifugation [21]. After the gradient is formed particles move within the gradient to the position having a density matching their own [22]. To improve our capacity to study mycobacteria in different cell states we describe a simple isopycnic technique to separate lipid rich and lipid poor mycobacteria based on their density. Our methodology was based upon isopycnic centrifugation with or without a centrifugation step [23]. This technique can produce very pure "single state" mycobacteria at good yield for use in further experimentation.

Another advantage of the method is that a solution of D_2O and pure water has no difficulties caused by evaporation. For other methods such as sucrose density centrifugation, a solution of sucrose and water will change its density if left uncovered overnight at room temperature due to the water evaporating off leaving comparatively more sucrose behind. Stability is another advantage as D_2O is atomically and there will be no change in solutes from precipitation mid-experiment caused by a change in the density of the media.

Confirmation and quality checking of the lipid-state of separated sub-populations can be obtained by use of the Nile red staining technique above and we report a simple method that assists the quantification of the LR and LP fractions.

To stain a bacterial culture or to grow it on differential and/or selective media is a standard and simple method for differen-

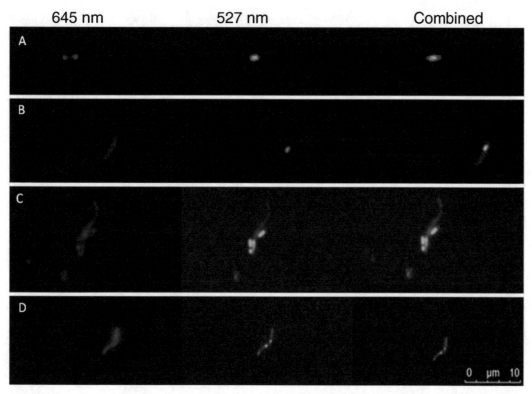

Fig. 1 Left to right; Nile red fluorescence of polar lipids at 645 nm, Nile red fluorescence of nonpolar lipids at 527 nm, composite image lipid rich cells extracted from D_2O separation top layer. (**a**) *M. marinum,* (**b**) BCG, (**c**) *M. smegmatis,* (**d**) *M. fortuitum.* It can be seen that in *M. marinum* the lipid body is single and large in the center of the cell. In BCG the lipid body (or bodies) is found at the polar end of the cell. *M. smegmatis* is similar to *M. marinum* in that it will have large lipid bodies situated in the center of the cell but *M. smegmatis* regularly displays more than one lipid body. *M. fortuitum* is similar again to BCG as it shows lipid bodies at the poles of the cells but much more commonly there will be two lipid bodies rather than one

tiating between genera of bacteria [24]. It is rapid and effective and remains an important part of everyday microbiology practice [24, 25]. Staining methodologies can also be adapted by adding more complex manipulations such as flow cytometry [26], cell counting and sorting and biomarker detection among other applications [27].

Nile red has been used for many years to visualize intracellular lipids. In this chapter we have adapted the methodology to provide a simple and reproducible technique to reliably visualize and quantify lipid bodies within mycobacterial cells. Importantly, the method can be adapted to flow cytometry. It is possible to further adapt our described Nile red staining protocol to a high throughput screening method to allow for rapid quantification of the lipid body load in a particular sample. Figure 1 below demonstrates what a lipid rich cell from a few mycobacterial species looks like and how they were identified microscopically.

When the preparation is visualized at 527 nm nonpolar lipids fluoresce green. Polar lipids such as those found in the cell wall and membranes appear as bright red. This clear visual separation allows for easy counting so that the proportion of lipid rich (LR) and lipid poor (LP) cells can be quantified accurately. This property can also be used by flow cytometry to rapidly quantify LR and LP cells in a mycobacterial culture [8] from a single staining step. This obviates the need for two stains: an acid-alcohol method to identify the mycobacterium followed by destaining followed by Nile red staining to classify the lipid content.

2   Materials

All solutions should be made prior to beginning this procedure. Diligently follow all waste disposal regulations when disposing of waste materials.

2.1   Buoyant Density Separation

1. Heavy water (Sigma-Aldrich).
2. Centrifuge (Beckman Coulter J6-MI).
3. Media (Sigma-Aldrich).
4. Microcentrifuge tubes (Axygen).
5. Pipettes (Thermo Scientific, Finnpipette F2).
6. Glass Pasteur pipette (Thermo Scientific).
7. Parafilm (Bemis, Parafilm).
8. Large centrifuge tubes (Cellstar).

2.2   Nile Red Staining

1. Nile red stain: Nile red power (Sigma-Aldrich) (*see* **Note 1**).
2. Clean microscope slides (Thermo Fisher).
3. DMSO (Sigma-Aldrich).
4. Microscope (*see* **Note 2**).
5. Water bath.

3   Method

3.1   Buoyant Density Separation (1 × g Separation)

1. Take a 1 mL aliquot of bacterial cells is harvested from culture.
2. Washed three times by microcentrifugation (20,000 × g for 3 min) with filter-sterilized water.
3. Resuspend the washed cells in 200 μL of filter-sterilized dH$_2$O.
4. The full 200 μL is aliquoted into an uncharged (or de-static) sterile plastic vessel (*see* **Note 3**).

5. Add 600 μL of D_2O to give final volume of 800 μL (D_2O:dH_2O; 3:1) (*see* **Notes 4** and **5**).

6. Seal the D_2O/dH_2O solution and leave to equilibrate for 24 h without agitation.

7. After 24 h take 100 μL of solution from within 1 mm of the meniscus using a 200 μL pipette.

8. Store the cells in a sterile microcentrifuge tube.

9. Remove the material from within 1 mm of the bottom of the tube with a 200 μL pipette (*see* **Note 6**).

10. Remove 100 μL from this layer and stored in a sterile microcentrifuge tube (*see* **Note 7**).

3.2 Microcentrifuge (~200 × g) Separation (See Note 8)

1. Cells for centrifugation were prepared as described in above.

2. Take an anti-static microcentrifuge tube is prepared.

3. Aliquot 1 mL mixture of washed cells in a 3:1 D_2O:H_2O into the tube.

4. Seal the tube and centrifuge for 5 min at 200 × *g*.

5. Take the fractions in same way as noted above in Subheading 3.1.

3.3 High Volume Preparation (200 × g) Separation (See Note 8)

1. Take a standard short nosed glass pipette and heat in a Bunsen burner whilst gripping the end of the pipette tip with forceps.

2. When the glass of the pipette tip begins to soften twist the forceps are and pulled to sever the end of the pipette tip and seal it in one movement.

3. Use this sealed and shortened pipette as the separation vessel.

4. Prepare a 5 mL solution of cells and 3:1 D_2O:H_2O is prepared as above and added to the sealed glass pipette.

5. Seal the pipette with Parafilm at the opening and place into a 15 mL centrifuge tube that has been padded with absorbent white tissue.

6. Ensure a good seal by adding more tissue paper around the pipette and above it before the cap of the large centrifuge tube is sealed.

7. Centrifuge the assembly at 200 × *g* for 5 min.

3.4 Staining Cells in Liquid Phase

1. Dissolve Nile red in DMSO (Sigma-Aldrich) to a final concentration of 2 mg/mL (*see* **Note 9**).

2. Nile red solution can be added directly into media containing cells at 1:10 ratio (final concentration of 100 μg/mL).

3. The Nile red sample should be incubated at room temperature with constant agitation for 20 min.

4. The sample is centrifuged for 3 min at 20,000 × *g* to pellet the cells.

5. The supernatant is removed and discarded and 200 μL of 100% ethanol is added. Vortex to mix.

6. The sample is centrifuged for 3 min at 20,000 × g and the supernatant discarded.

7. 100 μL of PBS is added to the sample and vortexed for 1 min.

8. The stained sample (10 μL) can be applied to a clean glass slide and heat-fixed.

9. Bacterial preparations can be viewed by fluorescence microscopy (*see* **Note 10**) or quantified by flow cytometry (*see* **Note 11**).

3.5 Staining Cells in Solid Phase (See Note 12)

1. Nile red is prepared as above to 100 μg/mL.

2. Sample bacterial cells using a sterile plastic loop.

3. Prepare a thin smear on a clean glass slide.

4. Heat-fix the smear.

5. Nile red bath is prepared with enough solution to flood the entire slide.

6. Place the prepared slide in a Nile red bath.

7. Bath and slide are incubated at room temperature for 30 min in the dark.

8. Remove the slide from bath and drain excess stain (*see* **Note 13**).

9. Slide has excess stain drained from it onto absorbent towelling.

10. Slide is rinsed once with deionized water, 3 s.

11. Slide is rinsed with 70% ethanol, 5 s.

12. Slide is rinsed again with deionized water and allowed to dry at room temperature in the dark (*see* **Note 12**).

4   Notes

1. Nile red is a benzophenoxazone dye and is highly soluble in ethanol but is negligibly soluble in water which makes its use in biological situations difficult. This can be overcome by bathing the sample to be stained in a highly polar substance. This can damage or change to properties of the sample under investigation so is generally avoided. An alternative is to use DMSO (dimethysulfoxide) as the solvent for the dye. Nile red is readily soluble in DMSO and DMSO will aid in the carriage of Nile red across biological membranes.

2. Any fluorescent microscope fitted with a 100× oil emersion lens and a >8 mega-pixel camera will suffice. The crucial elements that it must possess are filter cubes that fall within a fine

range. We use Texas red and Bodipy FL cubes as these have a narrow spectral range, ±40 nm of the stated wavelength.

3. This is achieved using an antistatic gun (Milty).

4. Deuterium oxide is a stable oxide of deuterium. Pure D_2O has a specific gravity of 1.11 g/cm^3. Pure water has a specific gravity of 1.00 g/cm^3. This means that a solution of D_2O from 1% to 99% could have the range of specific gravities from 1.01 to 1.10 g/cm^3. Previous work has shown that the density of lipid rich mycobacterial cells lies within this range (Lipworth, Gillespie, unpublished).

5. In order to effectively separate particles it is necessary to know the specific gravity of the particles in question. This can be established by performing several BDS' at a range of different specific gravities. It was found that lipid rich cells are separated at approximately 1.08 g/cm^3. Another cell type present in the sample (lipid poor cells) had a specific gravity of approximately 1.1 g/cm^3. In order to create a separation medium with a specific gravity similar to the density of lipid rich cells a mixture of D_2O and pure H_2O was used. Given the above figures (*see* Figs. 1 and 2) on the relative densities of pure H_2O (1.00 g/cm^3) and D_2O (1.11 g/cm^3) a 3:1 solution of $D_2O:H_2O$ gave a specific gravity of 1.08325 g/cm^3. This is slightly denser than the lipid rich cells under investigation. With a solution density of 3:1 $D_2O:H_2O$ a population of exclusively lipid rich cells gathered at the meniscus of the D_2O solution whereas all other cells sink to the bottom of the separation vessel.

6. Take the sample with bubbling through the D_2O/H_2O mixture until the correct depth was reached to prevent cells from other layers entering the pipette tip and contaminating the separated material.

7. When separations failed to achieve sufficient purity by fluorescent microscopic evaluation (*see* Fig. 1) such samples can be subjected to a further round of buoyant density separation.

8. For a microcentrifuge tube the maximum safe volume of liquid to be used is 1200 µL when centrifuging a sample. For a glass pipette, it is possible to use up to 5 mL of liquid.

9. Stain can be reused for subsequent staining up to a maximum of five times if stored in the absence of light or if used with 1 week of preparation.

10. For optimal clarity of separation we use an excitation frequencies of 480/40 and 540/40. We detect emission at 527/30, and 645/75. In our lab we use the Leica CTR 5500 DM microscope.

11. Preparations can be quantified using flow cytometry. Cells stained by the method noted as above in liquid phase are

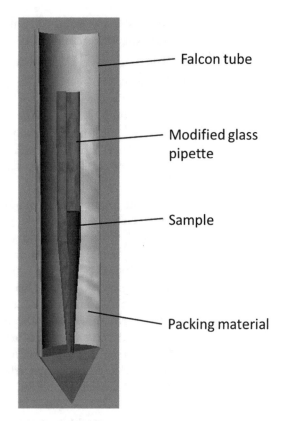

Fig. 2 Schematic diagram of the assembly used to centrifuge D_2O and bacterial samples for separation

loaded into flow cytometry vessel and processed as normal. Microscopic analysis of Nile red stained preparations is recommended to validate the results of flow cytometric analysis.

12. The same Nile red protocol that is used for solid phase cultures can be applied to ex vivo samples such as biopsy or postmortem sections. Slides must be dewaxed and distained if originally prepared thus. Ex vivo slides can then be processed as for solid phase, above.

13. Stain can be reused for subsequent staining up to a maximum of three times providing the whole of the sample remains on the slide.

References

1. Chen W, Zhang C, Song L, Sommerfeld M, Hu Q (2009) A high throughput Nile red method for quantitative measurement of neutral lipids in microalgae. J Microbiol Methods 77(1):41–47. https://doi.org/10.1016/j.mimet.2009.01.001

2. Feng GD, Zhang F, Cheng LH, Xu XH, Zhang L, Chen HL (2013) Evaluation of FT-IR and Nile Red methods for microalgal lipid characterization and biomass composition determination. Bioresour Technol 128:107–112. https://doi.org/10.1016/j.biortech.2012.09.123

3. Greenspan P, Mayer EP, Fowler SD (1985) Nile red: a selective fluorescent stain for intracellular lipid droplets. J Cell Biol 100(3):965–973

4. Parrish NM, Dick JD, Bishai WR (1998) Mechanisms of latency in Mycobacterium tuberculosis. Trends Microbiol 6(3):107–112

5. Dutta NK, Karakousis PC (2014) Latent tuberculosis infection: myths, models, and molecular mechanisms. Microbiol Mol Biol Rev 78(3):343–371. https://doi.org/10.1128/MMBR.00010-14

6. Deb C, Lee CM, Dubey VS, Daniel J, Abomoelak B, Sirakova TD, Pawar S, Rogers L, Kolattukudy PE (2009) A novel in vitro multiple-stress dormancy model for Mycobacterium tuberculosis generates a lipid-loaded, drug-tolerant, dormant pathogen. PLoS One 4(6):e6077. https://doi.org/10.1371/journal.pone.0006077

7. Baek SH, Li AH, Sassetti CM (2011) Metabolic regulation of mycobacterial growth and antibiotic sensitivity. PLoS Biol 9(5):e1001065. https://doi.org/10.1371/journal.pbio.1001065

8. Hammond RJH, Baron VO, Oravcova K, Lipworth S, Gillespie SH (2015) Phenotypic resistance in mycobacteria: is it because I am old or fat that I resist you? J Antimicrob Chemother. https://doi.org/10.1093/jac/dkv178

9. Schmidt WA (1919) Process and apparatus for separating finely-divided materials. Google Patents

10. Holter H, Ottesen M, Weber R (1953) Separation of cytoplasmic particles by centrifugation in a density-gradient. Experientia 9(9):346–348

11. Balentine R, Burford DD (1960) Differential density separation of cellular suspensions. Anal Biochem 1:263–268

12. Brakke MK, Daly JM (1965) Density-gradient centrifugation: non-ideal sedimentation and the interaction of major and minor components. Science 148(3668):387–389. https://doi.org/10.1126/science.148.3668.387

13. Boogaerts MA, Vercelotti G, Roelant C, Malbrain S, Verwilghen RL, Jacob HS (1986) Platelets augment granulocyte aggregation and cytotoxicity: uncovering of their effects by improved cell separation techniques using Percoll gradients. Scand J Haematol 37(3):229–236. https://doi.org/10.1111/j.1600-0609.1986.tb02302.x

14. Kurnick JT, Östberg L, Stegagno M, Kimura AK, Örn A, Sjöberg O (1979) A rapid method for the separation of functional lymphoid cell populations of human and animal origin on PVP-silica (Percoll) density gradients. Scand J Immunol 10(6):563–573. https://doi.org/10.1111/j.1365-3083.1979.tb01391.x

15. Makinoshima H, Nishimura A, Ishihama A (2002) Fractionation of Escherichia coli cell populations at different stages during growth transition to stationary phase. Mol Microbiol 43(2):269–279

16. Raposo G, Nijman HW, Stoorvogel W et al (1996) B lymphocytes secrete antigen-presenting vesicles. J Exp Med 183(3). https://doi.org/10.1084/jem.183.3.1161

17. Sartain MJ, Dick DL, Rithner CD, Crick DC, Belisle JT (2011) Lipidomic analyses of Mycobacterium tuberculosis based on accurate mass measurements and the novel "Mtb LipidDB". J Lipid Res 52(5):861–872. https://doi.org/10.1194/jlr.M010363

18. Daniel J, Maamar H, Deb C, Sirakova TD, Kolattukudy PE (2011) Mycobacterium tuberculosis uses host triacylglycerol to accumulate lipid droplets and acquires a dormancy-like phenotype in lipid-loaded macrophages. PLoS Pathog 7(6):e1002093. https://doi.org/10.1371/journal.ppat.1002093

19. Martin NJ, Macdonald RM (2015) Separation of non-filamentous micro-organisms from soil by density gradient centrifugation in Percoll. J Appl Bacteriol 51(2):243–251. https://doi.org/10.1111/j.1365-2672.1981.tb01238.x

20. Tsukamura M, Tsukamura S (1965) Grouping of mycobacteria by utilization of propylene glycol, glucose, fructose, and sucrose as sole carbon sources. Nihon Saikingaku Zasshi 20:229–232

21. Vincent R, Nadeau D (1984) Adjustment of the osmolality of Percoll for the isopycnic separation of cells and cell organelles. Anal Biochem 141(2):322–328

22. Arrowood MJ, Sterling CR (1987) Isolation of Cryptosporidium oocysts and sporozoites using discontinuous sucrose and isopycnic Percoll gradients. J Parasitol 73(2):314–319

23. Ishidate K, Creeger E, Zrike J, Deb S, Glauner B, MacAlister T, Rothfield L (1986) Isolation of differentiated membrane domains from Escherichia coli and Salmonella typhimurium, including a fraction containing attachment sites between the inner and outer membranes and the murein skeleton of the cell envelope. J Biol Chem 261(1):428–443

24. Leifson E (1951) Staining, shape and arrangement of bacterial flagella. J Bacteriol 62(4):377–389

25. Auty MAE, Gardiner GE, McBrearty SJ, O'Sullivan EO, Mulvihill DM, Collins JK, Fitzgerald GF, Stanton C, Ross RP (2001)

Direct in situ viability assessment of bacteria in probiotic dairy products using viability staining in conjunction with confocal scanning laser microscopy. Appl Environ Microbiol. https://doi.org/10.1128/AEM.67.1.420-425.2001

26. Nicoletti I, Migliorati G, Pagliacci MC, Grignani F, Riccardi C (1991) A rapid and simple method for measuring thymocyte apoptosis by propidium iodide staining and flow cytometry. J Immunol Methods 139(2):271–279

27. Amann RI, Binder BJ, Olson RJ, Chisholm SW, Devereux R, Stahl DA (1990) Combination of 16S rRNA-targeted oligonucleotide probes with flow cytometry for analyzing mixed microbial populations. Appl Environ Microbiol 56(6):1919–1925

Chapter 9

Methods to Determine Mutational Trajectories After Experimental Evolution of Antibiotic Resistance

Douglas L. Huseby and Diarmaid Hughes

Abstract

The evolution of bacterial resistance to antibiotics by mutation within the genome (as distinct from horizontal gene transfer of new material into a genome) could occur in a single step but is usually a multistep process. Resistance evolution can be studied in laboratory environments by serial passage of bacteria in liquid culture or on agar, with selection at constant, or varying, concentrations of drug. Whole genome sequencing can be used to make an initial analysis of the evolved mutants. The trajectory of evolution can be determined by sequence analysis of strains from intermediate steps in the evolution, complemented by phenotypic analysis of genetically reconstructed isogenic strains that recapitulate the intermediate steps in the evolution.

Key words Serial passage, Whole genome sequencing, Minimal inhibitory concentration, Relative fitness, Isogenic strains

1 Introduction

Antibiotics inhibit cell growth by targeting essential bacterial functions including the enzymatic activities of DNA gyrase, RNA polymerase, and the ribosome. Mutations in these enzymes that reduce the affinity of the drug for its target are one cause of resistance to antibiotics. Because these drug targets are highly conserved, and under strong selection for maximal functionality, most mutations that reduce drug-target interactions also reduce enzymatic functionality [1]. The reduced functionality associated with mutations causing antibiotic resistance sets up a dynamic situation where the mutant bacteria must continue to adapt by the acquisition of additional mutations. These additional mutations serve at least two different adaptive purposes: increasing resistance to the drug, and/or increasing relative fitness in the presence or absence of the selective drug. Examples of this type of multistep process are the evolution of resistance to fluoroquinolones in *E. coli* that requires the accumulation of several mutations to generate clinical resistance

Stephen H. Gillespie (ed.), *Antibiotic Resistance Protocols*, Methods in Molecular Biology, vol. 1736,
https://doi.org/10.1007/978-1-4939-7638-6_9, © Springer Science+Business Media, LLC 2018

[2, 3] and the evolution of rifampicin resistance in *M. tuberculosis*, caused by mutations in *rpoB*, that is frequently associated with secondary mutations in the same gene affecting resistance and/or relative fitness [4–7].

Experimental evolution of antibiotic resistance can be done in different ways: in liquid culture or on solid media; with constant or varying concentrations of antibiotic. The method we describe here is one we commonly use on our laboratory: evolving a bacterial strain in liquid culture, by serial passage at successively increasing concentrations of the drug, up to a predetermined drug concentration.

The exact trajectory of resistance evolution under selection by, for example, fluoroquinolones, cannot be accurately predicted a priori but must be confirmed after the evolution by whole genome sequencing (WGS) of the selected mutants. In experimental evolution, cultures or isolates taken after each of the different steps in the experiment should be stored so that they are available for later analysis. WGS is the method of choice to provide a complete genome-wide view into the genetic changes that have occurred. Typically WGS will be applied initially to strains or cultures from the starting point and the end point of the experiment. It may also be necessary to use WGS to analyze some or all of the stored strains/cultures from the different steps in the experiment so that the sequence of mutational events can be fully mapped. In this review we describe in outline how to do WGS to analyze experimental evolution experiments.

It is also necessary to understand the phenotypic changes that have accompanied the genetic changes revealed by WGS. In the evolution of antibiotic resistance the most important predicted change is that susceptibility to the selective drug should be reduced during the course of the experiment. While the selection pressure applied (in µg drug per mL culture) predicts a minimal level of susceptibility at each step in the experiment, the actual trajectory of susceptibility changes is usually nonlinear and must be measured for each mutant. The method of choice described in this review is to measure the minimal inhibitory concentration (MIC) of drug using a broth dilution protocol.

During the evolution of resistance the mutations that accumulate in the selected strains will frequently be the cause of changes in relative fitness. Methods to measure relative fitness, as a function of antibiotic concentration under laboratory conditions, have been recently described [8, 9].

2 Materials

2.1 Serial Transfer Components

1. Mueller-Hinton II broth (MHII): cation-adjusted (from Becton Dickinson, cat. no. 212322) for bacterial growth, and

for culture and drug dilutions throughout the assay. Make up according to the manufacturer's instructions.

2. Mueller-Hinton II agar: (from Becton Dickinson, cat. no. 211441) for bacterial growth. Make up according to the manufacturer's instructions.

2.2 Whole Genome Sequencing Components

1. It is not possible to give details here (there are too many details and different options available) other than to state that there are several commercially available kits for the preparation of genomic DNA, and for its preparation into libraries suitable for whole genome sequencing. Currently popular sequencing technologies are marketed under the Illumina and PacBio brands, but there are others. Many users will probably make use of in house genome sequencing facilities, or commercial companies that perform all of the steps and also provide bioinformatics analysis services.

2.3 Minimal Inhibitory Concentration (MIC) Components

1. Mueller-Hinton II broth (MHII): cation-adjusted (Becton Dickinson, cat. no. 212322) for bacterial growth, and for culture and drug dilutions throughout the assay. Make up according to the manufacturer's instructions.

2. 96-well (12 × 8) microtiter plates: use round-bottomed plates (VWR, cat. no. 732-2725).

3. Sterile polyester adhesive film for microplates (VWR, cat. no. 60941-064).

4. Microwell lid for 96-well plates (VWR, cat. no. 734-2185).

3 Methods

3.1 Experimental Evolution by Serial Transfer

The goal of serial transfer experiments is to recreate selections that bacteria may face that lead to the development of antibiotic resistance. To this end, the choice of selective conditions is absolutely essential. Below is described a liquid-media selection in which the concentration of the fluoroquinolone antibiotic ciprofloxacin is increased in 1.5-fold steps. Practically, the described experiment leads to the development of *E. coli* mutants that are clinically resistant to ciprofloxacin (EUCAST breakpoint of 1 μg/mL) after 14 cycles.

Depending on the experimental question being addressed, different selection regimes will be called for. Liquid serial transfer experiments, similar to the one described below, impose a selection pressure for fast growth in addition to antibiotic resistance. Similar experiments done by plating for single colonies on solid media containing antibiotics will, in contrast, select for resistance almost exclusively.

1. From a frozen stock, streak out wild-type *E. coli* MG1655 for single colonies on a Muller-Hinton agar plate. Incubate overnight at 37 °C (*see* **Note 1**).

2. Begin ten separate 2 mL liquid overnight cultures in Muller-Hinton media in 10 mL disposable Falcon tubes. Use a separate colony to inoculate each tube. Incubate the tubes for 24 h at 37 °C with vigorous aeration (200 rpm) (*see* **Notes 2** and **3**).

3. Transfer 900 μL of each overnight culture to 2 mL screw-cap tubes containing 225 μL of sterile 50% glycerol in Muller-Hinton II Broth. These tubes should be stored at −80 °C for subsequent analysis.

4. Transfer 2 μL of each culture into a new tube containing 2 mL of fresh Muller-Hinton media with 0.016 μg/mL ciprofloxacin (a concentration corresponding to the MIC of *E. coli* MG1655). Incubate the tubes at 37 °C for 24 h with vigorous aeration (*see* **Notes 4** and **5**).

5. Transfer 2 μL of each culture into a new tube containing 2 mL of fresh Muller-Hinton media with 0.024 μg/mL ciprofloxacin (a concentration corresponding to 1.5× the MIC of *E. coli* MG1655). Incubate the tubes at 37 °C for 24 h with vigorous aeration (*see* **Notes 4** and **5**).

6. Transfer 900 μL of each overnight culture to 2 mL screw-cap tubes containing 225 μL of sterile 50% glycerol in Muller-Hinton II Broth. These tubes should be stored at −80 °C for subsequent analysis.

7. Repeat **steps 5** and **6**, increasing the concentration of antibiotic 1.5-fold at each cycle. Continue until the desired level of resistance is achieved.

8. Store the final culture as a frozen stock at −80 °C for subsequent analysis.

3.2 Whole Genome Sequencing

Whole genome sequence analysis is the method of choice to determine the mutational events that occurred during the selected evolution of resistance to an antibiotic.

1. Cultures from the frozen stocks from the final serial transfer step (Subheading 3.1, **step 8**) are struck out for single colonies and incubated overnight at 37 °C on Mueller-Hinton II agar plates containing the same concentration of ciprofloxacin as was used in the final step of selection (*see* **Note 6**).

2. Single colonies are picked and used to inoculate independent tubes containing 2 mL of Mueller-Hinton II broth supplemented again with the same concentration of ciprofloxacin. These tubes are incubated for 16–24 h at 37 °C with vigorous aeration (200 rpm).

3. Transfer 900 μL of each overnight culture to 2 mL screw-cap tubes containing 225 μL of sterile 50% glycerol in Muller-Hinton II broth. These tubes should be stored at −80 °C (*see* **Note 7**).

4. The remaining culture should be used to prepare genomic DNA. A wide range of commercial kits is available for the preparation of genomic DNA.

5. The genomic DNA should be whole-genome sequenced. There are many options for whole-genome sequencing. Most groups have their own sequencing capabilities, use an institutional core facility, or take advantage of one of a number of commercial options (*see* **Note 8**).

6. The resultant sequence should be analyzed versus a reference sequence for SNPs, deletions, insertions, and copy-number variants.

7. It is likely that multiple mutations will have accumulated, raising at least two questions that will need to be addressed. (a) What is the order in which individual mutations occurred? (b) What is the nature of the contribution of individual mutations to the selected phenotype?

8. Question (a), the order in which mutations accumulated, can be addressed by repeating WGS **steps 1–6** on clones isolated from the stocks made of earlier steps in the evolutionary serial passage experiment.

9. Question (b) can be addressed by using genetics to reconstruct isogenic strains carrying suitable combinations of mutations identified from the WGS analysis and then measuring their MIC (described in Subheading 3.3 below) and measure their relative fitness under an appropriate condition [8, 9].

10. Relating MIC and relative fitness with the data from WGS is expected to give a useful insight into the nature of the selective advantage conferred by each successive mutation that accumulated during the experimental evolution experiment.

3.3 MIC Assay by Broth Dilution

The minimal inhibitory concentration (MIC) of an antibiotic is defined as the lowest concentration of drug that, under a defined set of agreed conditions, prevents visible growth of a bacterial culture. The procedure that follows outlines the conditions that should be met and the procedures that should be followed when using broth dilution, in microtiter plate format, to measure MIC. The conditions follow closely those recommended by EUCAST [10] and are applicable to *E. coli* and fluoroquinolones, but details may need to be changed depending on the growth requirements of particular bacterial species and the properties of particular antibiotics being tested.

1. Plate format 8 × 12 (rows A–H, and columns 1–12). The formatting of strains versus antibiotic concentrations can be planned according to the requirements of the particular project. We typically set up the assay so that on one microtiter plate we test eight strains against 11 concentrations of drug (including zero), and 1 medium sterility control.

2. A stock solution of the antibiotic compound should be made as recommended by the manufacturer and diluted in MHII medium to twice the highest required final concentration. In the example given here, a ciprofloxacin stock solution (1 mg/mL in 0.1 N HCl) is diluted to 64 μg/mL in MHII medium.

3. Drug assay concentrations: Drug concentrations are usually varied in twofold dilution steps. The drug concentration range should be set to match the properties of the particular experimental expectations for the bacteria-drug combination being studied. For example, in the fluoroquinolone resistance evolution experiment a typical range might be as follows, in μg/mL: 32: 16: 8: 4: 2: 1: 0.5: 0.25: 0.12: 0.06: 0.00: and 0.00 (medium only sterility control).

4. Transfer 5–10 mL sterile MHII medium into a sterile 50 mL reagent reservoir (Corning Incorporated, cat no. 4870).

5. Medium inoculation into microtiter plate: Using a multichannel pipette (eight channels) inoculate 50 μL MHII medium from the reagent reservoir into the wells of columns 2–11.

6. Inoculate 100 μL MHII medium into the wells of column 12 (this will be the medium sterility control).

7. Drug inoculation into microtiter plate: Inoculate 100 μL of the 64 μg/mL ciprofloxacin solution into each of the column 1 wells.

8. Using a multichannel pipette (eight channels) transfer 50 μL of the ciprofloxacin solution from column 1 into the column 2 wells.

9. Repeat **step 9**, successively transferring 50 μL from column 2 to column 3, and so on until column 10.

10. Using a multichannel pipette (eight channels) remove 50 μL of the ciprofloxacin solution from column 10 and discard it as waste.

11. At this stage columns 1–10 have 50 μL of MHII with drug, column 11 has 50 μL MHII without drug, and column 12 has 100 μL MHII without drug.

12. Bacterial inoculum preparation: Suspend fresh colonies of each strain of interest, grown on nonselective medium (incubation 18–24 h at 35 °C ± 2 °C) in saline (0.9% NaCl) to 0.5 McFarland ($\cong 1.5 \times 10^8$ CFU/mL) (*see* **Note 9**).

13. Transfer 50 μL of each bacteria suspension to 10 mL of MHII broth to make a set of final bacterial suspensions: $\cong 5 \times 10^5$ CFU/mL (acceptable range $3–7 \times 10^5$ CFU/mL).

14. Using a 12 channel multichannel pipette (using only 11 of the 12 channels) transfer 50 μL of a bacterial strain to column A (wells 1–11). Well 12 is a medium control. The assay volume in each well in 100 μL.

15. Repeat **step 14** for each of the seven remaining bacterial strains. If more strains are to be tested, apply the same procedure to additional microtiter plates as appropriate.

16. Cover the plates with a sterile polyester adhesive film for microplates (VWR, cat. no. 60941-064), and a microwell lid for 96-well plates (VWR, cat. no. 734-2185).

17. Incubate without shaking for 16–20 h at 35 °C ± 2 °C. Do not stack the plates more than four high (to maintain the same incubation temperature for all plates).

18. Read MIC visually. MIC is defined as complete inhibition of growth as detected by unaided eye, using medium only as the control.

19. Each strain or antibiotic should be assayed in duplicate (independent plates) to test the robustness of the data.

4   Notes

1. BBL Mueller-Hinton II Agar (BD) plates are prepared according to manufacturer instructions.

2. BBL Mueller-Hinton II Broth (BD) is prepared according to manufacturer instructions. Once the media has cooled following autoclaving, antibiotics can be added. Bottles with media supplemented with antibiotics may be stored at 4 °C during the course of selection until required. It is helpful to preprepare many small bottles containing 100–200 mL of media supplemented with the various concentrations of antibiotic you intend to use during the whole course of the serial passage experiment.

3. Ten parallel cultures are routinely used as a standard scale of experiment. Using smaller numbers of replicates increases the possibility that the full spectrum of mutational paths toward resistance will not be observed, while increased numbers of replicates increases the complexity and expense of the experiment, particularly if multiple selection regimes are investigated simultaneously.

4. For some antibiotics and concentrations it is possible that 24 h of growth will be insufficient to achieve full density cultures

($>10^9$ cells/mL for *E. coli*). In this case, either longer incubations or smaller antibiotic concentration increases between serial transfers may be beneficial. Growth from a 1000-fold dilution, as described above, to full density corresponds to ten generations of growth.

5. Maintaining proper sterile technique is absolutely essential, since cross-contamination of lineages may lead to less information about potential resistance development trajectories. Blank, un-inoculated media control tubes should always be prepared and incubated in parallel with experimental tubes. Regular spraying and wiping down of equipment and workspaces with 70% ethanol is advised. Use of a laminar flow cabinet, if available, is also beneficial.

6. Resistance to antibiotics is often conferred fully or partially by genetic amplification of genes on the chromosome or plasmids. In the absence of selection, these amplifications are rapidly lost from the population. In order to detect these amplifications in sequencing data, it is important that they are always held by selection during all the growth steps for genomic DNA preparation.

7. Evolutions in liquid culture will to varying degrees generate mixed populations. Generally sequencing single clones from a population is preferred to sequencing the whole population. Mixed population sequencing may be possible depending on the number of genotypes represented in the population and the sequencing technology used, but also may present problems both predictable and unexpected. If a single-clone sequencing strategy is used, it is essential that the specific clone that is sequenced be saved. Reisolating a particular clone from a population can be labor-intensive and success cannot be guaranteed!

8. The most popular whole genome sequencing technology currently for this type of analysis is that sold under the Illumina brand. Illumina sequencing technology generates large numbers of short, paired reads (~75–300 bp). This type of data is very useful in cases where a high-quality reference sequence is available for the organism being sequenced. In this case Illumina data can be used to find SNPs, short insertions and deletions, and copy-number variants. In the event that the genome being sequenced contains many repetitive sequences it may be appropriate to use another technology, alone or in combination with Illumina, that can generate longer contiguous reads. In such instances a popular current technology is that marketed under the PacBio brand.

9. For each species there are recommended quality control strains [10] that can be purchased from the American Type Culture Collection and these should be routinely used to ensure that the conditions of the MIC assay are within acceptable margins.

Acknowledgments

This work is supported by grants from the Swedish Research Council (VR-NT and VR-Med (not VE as written)-Med), Swedish Council for Strategic Research (SSF), and the Knut and Alice Wallenberg Foundation (RiboCORE project) to D.H.

References

1. Andersson DI, Hughes D (2010) Antibiotic resistance and its cost: is it possible to reverse resistance? Nat Rev Microbiol 8(4):260–271

2. Marcusson LL, Frimodt-Moller N, Hughes D (2009) Interplay in the selection of fluoroquinolone resistance and bacterial fitness. PLoS Pathog 5(8):e1000541

3. Komp Lindgren P, Karlsson Å, Hughes D (2003) Mutation rate and evolution of fluoroquinolone resistance in Escherichia coli isolates from patients with urinary tract infections. Antimicrob Agents Chemother 47(10):3222–3232

4. Brandis G et al (2014) Comprehensive phenotypic characterization of rifampicin resistance mutations in Salmonella provides insight into the evolution of resistance in Mycobacterium tuberculosis. J Antimicrob Chemother 70(3):680–685

5. Casali N et al (2012) Microevolution of extensively drug-resistant tuberculosis in Russia. Genome Res 22(4):735–745

6. Comas I et al (2012) Whole-genome sequencing of rifampicin-resistant Mycobacterium tuberculosis strains identifies compensatory mutations in RNA polymerase genes. Nat Genet 44(1):106–110

7. Brandis G et al (2012) Fitness-compensatory mutations in rifampicin-resistant RNA polymerase. Mol Microbiol 85(1):142–151

8. Huseby DL, Pietsch F, Brandis G, Garoff L, Tegehall A, Hughes D (2017) Mutation supply and relative fitness shape the genotypes of ciprofloxacin-resistant Escherichia coli. Mol Biol Evol 34(5):1029–1039

9. Praski Alzrigat L, Huseby DL, Brandis G, Hughes D (2017) Fitness cost constrains the spectrum of marR mutations in ciprofloxacin-resistant Escherichia coli. J Antimicrob Chemother 72(11):3016–3024

10. European Society of Clinical Microbiology and Infectious Diseases (2013) Antimicrobial susceptibility testing. http://www.eucast.org/

Chapter 10

Selection of ESBL-Producing *E. coli* in a Mouse Intestinal Colonization Model

Frederik Boëtius Hertz, Karen Leth Nielsen, and Niels Frimodt-Møller

Abstract

Asymptomatic human carriage of antimicrobially drug-resistant pathogens prior to infection is increasing worldwide. Further investigation into the role of this fecal reservoir is important for combatting the increasing antimicrobial resistance problems. Additionally, the damage on the intestinal microflora due to antimicrobial treatment is still not fully understood. Animal models are powerful tools to investigate bacterial colonization subsequent to antibiotic treatment. In this chapter we present a mouse-intestinal colonization model designed to investigate how antibiotics select for an ESBL-producing *E. coli* isolate. The model can be used to study how antibiotics with varying effect on the intestinal flora promote the establishment of the multidrug-resistant *E. coli*. Colonization is successfully investigated by sampling and culturing stool during the days following administration of antibiotics. Following culturing, a precise identification of the bacterial strain found in mice feces is applied to ensure that the isolate found is in fact identical to the strain used for inoculation. For this purpose random amplified of polymorphic DNA (RAPD) PCR specifically developed for *E. coli* is applied. This method allows us to distinguish *E. coli* with more than 99.95% genome similarity using a duplex PCR method.

Key words Extended-spectrum beta-lactamase (ESBL), *E. coli*, Mouse model, Intestinal colonization, RAPD, Typing, Selection, Antibiotics, Antibiotic resistance

1 Introduction

A major source for antimicrobial resistance in *E. coli* is plasmid-borne *Extended-Spectrum β-Lactamases* (ESBL) [1–4]. The majority of ESBLs belong to the four large families of SHV, TEM, CTX-M and OXA [2, 5]. ESBL CTX-M enzymes were first described in Germany in 1989, but rapidly it became the dominating ESBL genotype during the early 2000s [6]. The rapid worldwide dissemination has been known as the "CTX-M pandemic" and the dominance of CTX-M types has largely been caused by dissemination of *E. coli* lineages [7–9]. Now, community-onset ESBL infections have become an important public health issue, as community-onset infections caused by ESBL-producing *E. coli* primarily are caused by CTX-M ESBLs [6, 9, 10].

Stephen H. Gillespie (ed.), *Antibiotic Resistance Protocols*, Methods in Molecular Biology, vol. 1736,
https://doi.org/10.1007/978-1-4939-7638-6_10, © Springer Science+Business Media, LLC 2018

The indigenous gastrointestinal microflora acts as a barrier against incoming pathogens and overgrowth of opportunistic microorganisms already present in the gut. Alterations in the microbiota can create a window for opportunistic pathogens leading to possible overgrowth of resistant strains [11–15]. One of the most dramatic modifications to the gut community is antibiotic treatment [11, 12]. Overgrowth and establishment of resistant strains is not seen to the same degree in individuals not under the influence of antibiotic [16–19]. An association of antibiotics and presence of resistance has been described [16–20].

Mice are used extensively in animal experiments to study the impact on intestinal colonization of different bacterial species with relatively low cost and with good reproducibility [21, 22]. Furthermore, the intestinal flora of laboratory mice and men are comparable, which is why mice are often the first mammal used to explore the association between intestinal microbiome, health and disease [23]. Mouse models are also used to investigate bacterial colonization and studies often include the administration of antibiotics *prior* to inoculation of the bacteria of interest [23–29].

Mouse intestinal colonization of Gram-negative bacteria has successfully been determined in several studies. Yet most models use elimination of resident facultative bacteria prior to inoculation [24, 26, 30–32]. As such, mice treated with streptomycin have been shown to be susceptible to enteric infection [11]. Additionally, previous mice-studies have shown that exposure to sub-therapeutic concentrations of penicillin, vancomycin, penicillin and vancomycin, or chlortetracycline produced elevated ratios of *Firmicutes* to *Bacteroidetes*. In addition, treatment with antibiotics of broad spectrum of activity, or impact on the anaerobic flora, has been shown to reduce the *Bacteroidetes* population [11, 12, 31]. Furthermore, Perez et al. studied the effect of subcutaneous treatment on the indigenous intestinal microflora of mice and investigated the effect on colonization by a KPC-producing *Klebsiella* strain (KPC-Kp strain) [31]. Their findings indicated that the anaerobic effect of antibiotics promoted the establishment of the KPC-Kp strain—provided that the antibiotic had no effect on the strain carrying resistance genes. Thus, antibiotics with limited effect on the anaerobic flora might be less likely to promote the colonization of multi-drug resistant Gram-negative bacteria [31].

Enterobacteriaceae, such as *E. coli* are able to colonize and survive in many different locations, including the human gastrointestinal tract [9, 33]. Moreover, *E. coli* is one of the organisms most frequently found harboring genes coding for Extended-spectrum beta-lactamases (ESBL) [3, 8, 34]. In recent years asymptomatic carriage of antimicrobially resistant *E. coli* in humans has been found in different parts of the world, with low to modest carrier rates of 6–13% and high carrier rates of 50–63% [35–38].

The indigenous intestinal flora acts as a barrier against incoming pathogens and overgrowth of opportunistic microorganisms already present in the gut; known as colonization resistance. However, alterations in the microbiota can allow for colonization, with possible subsequent infection, and antibiotic treatment is known to disturb the ecological balance of the indigenous microflora [13–15].

Understanding how antibiotics select for an ESBL-producing *E. coli* isolate is very important and here we present a mouse-intestinal colonization model designed to explore this phenomenon. The model can be used to study how antibiotics with varying effect on the intestinal flora promote the establishment of the multi-drug resistant *E. coli*—provided the antibiotic has no effect on the administered bacterial strain harboring the relevant resistance mechanism. Colonization is successfully investigated by sampling and culturing stool during the days following administration of antibiotics [29]. A precise identification of the bacterial strain found in mice feces is of great importance to ensure that the isolate found is in fact identical to the strain used for inoculation. For this purpose random amplified of polymorphic DNA specifically developed for *E. coli* is applied. This method allows us to distinguish *E. coli* with more than 99.95% genome similarity using a duplex PCR method, based on short primers binding randomly to the DNA, resulting in fingerprints of each strain on a gel [39].

2 Materials

1. Chromogenic agar plates: Discovery agar base (Oxoid) containing 32 μg/mL cefotaxime (*Fresenius-Kabi*) and 6 μg/mL vancomycin (Sigma-Aldrich) (*see* **Note 1**).

2. 5% Blood-agar plates containing 4 μg/mL gentamicin (SSI Diagnostica) (*see* **Note 2**).

3. Anaerobic agar plates containing 32 μg/mL gentamicin and 16 μg/mL vancomycin (SSI Diagnostica) (*see* **Note 3**).

4. 0.9% saline: MilliQ water, 0.9% NaCl.

5. Freezing stock: Luria Broth (Sigma-Aldrich), 15% glycerol (Sigma-Aldrich).

6. Disposable needles (BD Medical).

7. Syringes (BD Medical).

8. Female albino outbred NMRI mice: 7–10 weeks old, weighing 26–30 g (Harlan) (*see* **Note 5**).

9. Disposable, sterile tweezers (Unomedical).

10. Wide-bore 50–1000 μL pipette tips (Sartorius).

11. E-Gel® Electrophoresis System (Thermo Scientific).

12. Gel electrophoresis UV camera (Bio-Rad).

13. A colonizing *E. coli*: Here, A clinical blood isolate of *E. coli*, lineage B2-O25b-ST131 (*see* **Note 8**).

14. Colorimeter 254, 546 nm (Sherwood).

15. Multiplex PCR kit (Qiagen).

16. Primers 1283 (GCGATCCCCA) (TAG Copenhagen).

17. Primer 1247 (AAGAGCCCGT) (TAG Copenhagen).

18. 2% E-gels, 48 wells (Invitrogen).

19. 1 kb DNA ladder (Fermentas).

20. Positive control for RAPD PCR: *E. coli* ATCC 25922 (Oxoid).

21. A PCR cycler (e.g., Applied Biosystems).

22. Antibiotics for treatment of the mice, dissolved in 0.9% saline to appropriate concentration.

23. 2 g Ampicillin (Bristol-Myers Squibb,), daily dose: 1.5 mg/mouse.

24. 1 g Cefotaxime (Fresenius-Kabi), daily dose: 2 mg/mouse.

25. 1.5 g Cefuroxime (Fresenius-Kabi), daily dose: 4 mg/mouse.

26. 1 g Meropenem (Hospira), daily dose: 1.5 mg/mouse.

27. 2 mg/mL Ciprofloxacin (Fresenius-Kabi), daily dose: 0.5 mg/mouse.

28. 1 g Diclocil (Bristol-Myers), daily dose: 2 mg/mouse.

29. 1 g Selexid (Leo-Pharma), daily dose: 2 mg/mouse.

30. 150 mg/mL Clindamycin (Stragen), daily dose: 1.4 mg/mouse.

31. 5 mg/mL Metronidazole (Baxter), daily dose: 2.5 mg/mouse.

Agar containing antibiotics should be stored at 4–5 °C and kept in darkness to avoid untimely decomposition of antibiotics.

3   Methods

3.1 Mouse Experiments

1. Mice are brought into the stable 4–7 days prior to investigation. In the stable they are divided into pairs of two per cage and two cages belong to one group; thus, one antibiotic is given to a total of four mice. The study is conducted from day 1 to day 8. All cages are changed daily. At the end of day 8 all mice are sacrificed (*see* **Notes 4–7**).

2. On day zero, culture the colonizing pathogen (CP) on suitable agar plates (*see* **Notes 8** and **9**).

3. On day 1

(a) Suspend the CP in 0.9% saline to 10^8 CFU (colony forming units)/mL using a colorimeter (*see* **Note 10**).

(b) Administer the suspension to the mice as 0.25 mL through a stainless steel orogastric feeding tube.

(c) Leave the mice subsequently for 3 h, before the first dose of antibiotic is administered subcutaneously as a maximum of 0.25 mL as according to good practice [31, 40] (*see* **Notes 11** and **12**). To effectively mimic the serum antibiotic concentrations achieved in humans (on standard doses), all mouse doses for antibiotic administration are calculated based on human dose (in mg/kg of body weight) from previously published mouse-studies or PK-studies performed at Statens Serum Institut [22, 29, 31, 41–45] (*see* **Note 13**).

(d) Mice droppings are collected into individual 15 mL Nunc™ tubes for each cage (feces belongs to two mice) using disposable and sterile tweezers (*see* **Note 14**). Collect a total of 0.5 g of feces from each cage.

4. On day 2 and 4

(a) The mice inhabiting the cage are moved to a clean cage preceding sampling of feces.

(b) Individual mice droppings are collected into a 15 mL Nunc™ tube as described on day 1.

(c) The antibiotics are administered subcutaneously.

5. On day 8

(a) The mice are sacrificed.

(b) Individual mice droppings are collected into a 15 mL Nunc™ tube as described on day 1.

3.2 Determining Colony Forming Units

1. Three different plates are used for identification of ESBL-producing *E. coli*, Gram-positive bacteria and *Bacteroides*, respectively (*see* **Notes 1–3**).

2. On the day of collection (preferentially done within 1 h of sampling), dissolve the 0.5 g of individual feces samples in 5 mL of saline, vortex and leave for 1 h.

3. Dilute the samples tenfold in saline by serial passage of 100 μL sample to 900 μL of saline a total of six times (dilutions 10^{-1} to 10^{-6}). 20 μL of each solution is spotted on duplicate plates of each of the different selective agars (six plates in total)—creating a circle of spots with the undiluted solution placed in the middle (*see* **Note 15**). The spots cannot touch each other. Leave the plates to dry on the table until the spot is completely dry before moving to culturing. Culture the plates under appropriate conditions for 18–20 h (37 °C) (*see* **Notes 1–3**).

4. After 18–20 h, calculate the CFU number for the resistant strain in question, the anaerobic Gram-negative flora and the aerobic gram-positive flora in each sample from the number of colonies on the plates.

From the selective plates calculate the LOG CFU number per 0.5 g of stools for the three bacterial populations for each antibiotic and the control groups. Lower detection limit is LOG(50) per 0.5 g of feces (*see* **Note 16**).

3.3 Storage of Isolates and RAPD

1. On day 1, 2, 4, and 8 pick several randomly selected *E. coli* colonies from plates used for CFU calculations. Streak the chosen colonies on to blood agar, incubate overnight and freeze a loopful (5 μL inoculation needle) in broth containing 15% glycerol at −80 °C. This is done for plates representing each cage.

2. Crude Extract DNA lysates are made by taking 1 μL (1 μL inoculation needle) of bacteria from a plate and transfer it into 300 μL of DNase-free water (Invitrogen) and incubate for 10 min at 95 °C. Centrifuge at $5000 \times g$ for 10 min to pellet cell remnants. Transfer the supernatant to a new tube or avoid touching the pellet when using the DNA. Store at −20 °C.

3. Dilute the primers to 20 μM in DNase-free water (*see* **Note 17**).

4. PCR mastermix: Two different PCR mastermixes are created, each containing one of the two primers. For each sample mix 12.5 μL of multiplex PCR kit without Q-solution (Qiagen), 7.5 μL of DNase free water, and 2.5 μL of 20 μM primer (1247 or 1283). Create a batch mix when testing more samples and distribute into PCR tubes afterward. Then, add 2.5 μL of template DNA (or DNase-free water as negative control) to each tube and start the appropriate reaction in the thermal cycler:

   (a) 95 °C for 15 min.

   (b) 35 cycles of 94 °C for 1 min, 38 °C (primer 1247)/36 °C (Primer 1283) for 1 min, 72 °C for 2 min.

   (c) 10 min elongation step at 72 °C.

5. Electrophoresis:

   (a) The E-Gel® is placed in the holder of the E-Gel® electrophoresis system.

   (b) Load 15 μL of DNase-free water and 5 μL of PCR product for each sample.

   (c) The marker is loaded as 1 μL of marker to 9 μL of DNase-free water.

   (d) The gel is run for 34 min at 50 V.

(e) Take a picture of the gel by your local UV gel electrophoresis camera.

(f) The fingerprinting will differ in number of bands for different *E. coli* clones (less than 99.5% genome similarity) (*see* **Note 18**).

---

4 Notes

1. The Chromogenic agar used can identify and quantify uropathogenic bacteria according to color (e.g., *E. coli* are red, *Klebsiella* spp. are blue, and *Enterococcus* spp. are green/blue). These agar plates are used to identify the colonizing pathogen and should therefore contain antibiotics with effect on the indigenous microbiota of the mouse, but without effect on the colonizing pathogen (the ESBL-producing *E. coli*).

2. Blood agar plates with gentamycin inhibit growth of any susceptible *E. coli* and are therefore used to evaluate antibiotic impact on the Gram-positive flora of the mice gut. Gentamicin is used since the colonizing ESBL-producing *E. coli* is susceptible to gentamicin.

3. The anaerobic agar plates are used to evaluate the impact of antibiotics on the population of *Bacteroides*. These plates must be cultured under anaerobic conditions. If no anaerobic chamber is available, culturing of anaerobic spp. can be performed in sealed jars with anaerobic atmosphere, as created in GASPAK EZ containers by AnaeroGen® (Oxoid) [46].

4. We performed all experiment at Statens Serum Institut (SSI) in Copenhagen, Denmark. All animal experiments must be approved by the local Centre for Animal Welfare and carried out at approved facilities by trained personal with required certificates.

5. All mice should be from the same litter.

6. No alterations of the intestinal flora are induced prior to the study, i.e., no antimicrobial treatment is administered.

7. It is advised to perform experimental studies to ensure that the specific mouse-intestinal colonization model produces reliable and reproducible results. For our experimental studies, we used nine groups, with each group receiving one antibiotic only. Each group consisted of two cages with two mice each. For all studies, we included a control group.

8. The mouse intestinal colonization can be applied on isolates with other resistance mechanisms than ESBL. This requires that the applied colonizing isolate is resistant to the drugs administered and that the isolate chosen as colonizer, must be

resistant to one of the antibiotics used in the chromogenic agar and susceptible to the antibiotic used in blood agar and anaerobic agar plates. We suggest that phenotypic and genotypic characterization of isolates is performed prior to investigation, to ensure easy and effective identification of isolate in feces, e.g., MAST-test, MLST, and RAPD. RAPD is used for successive identification. For a similar study it would be possible to use other ESBL- or carbapenemase-producing *E. coli* of other ST's or serogroups; such as B2-O16-ST16, ST69 or ST153.

9. Prior to investigation all plates must be tested. At least three chromogenic agar plates from each batch are tested with one susceptible and one resistant *Enterobacteriaceae* and finally one vancomycin-susceptible *Enterococcus* spp. Store at 4–5 °C and keep out of light. Alternatively cover them with tinfoil.

10. CFU counts of CP suspensions are performed as described for the fecal samples to verify the inoculum.

11. A single daily subcutaneous dose can produce similar levels of drugs in mice feces, to those seen in humans [31]. Saline solution containing antibiotics should be prepared on a daily basis to ensure a precise and correct concentration. Information on dilution and storage should be done according to instructions from manufacturer and further guidance can be found from current literature, such as Andrews et al. [47]. Solutions should be kept in dark or covered bottles to avoid disintegration and change in concentration. All antibiotics are dissolved in sterile 0.9% saline to avoid any pain or wounding of mice during administration.

12. We include a control group receiving the CP but no antibiotic treatment and one group receiving treatment with cefotaxime with no inoculation with CP.

13. For this study we used ampicillin, cefotaxime, cefuroxime, ciprofloxacin, meropenem, dicloxacillin, mecillinam, clindamycin, and metronidazole, but other antibiotics can be applied.

14. New sets of tweezers are used for each cage.

15. Between each serial passage the pipette tip should be changed and the solution vortexed. The plates used for the spot CFU method should be completely dry. If this is not the case the spots will mix with each other. Dry your plates beforehand in an incubator.

16. Example of CFU calculations:
    CFU spots are performed on duplicate plates using the same dilution row. Colonies are therefore counted in two spots for each dilution on each type of plate. The samples are diluted tenfold (0.1 mL of sample diluted in 0.9 mL of 0.9% saline). Thus, first spot is undiluted, the second is diluted ten-

fold (10^{-1}), the third yet again diluted tenfold (10^{-2}), and so on (10^{-3} to 10^{-6}).

Spot-calculation: Average of the duplicate counts × 50 (because we plated 20 μL, 1/50 of 1 mL) × dilution factor (10^{-1} to 10^{-6}). That will yield the number CFU/mL.

17. Primers can be batch diluted to 20 μM and then aliquoted into separate Eppendorf tubes and frozen at −20 °C. Then you only defrost your primers once before usage.

18. For *E. coli* RAPD patterns differing by >1 band difference are considered different clones.

References

1. Rogers BA, Sidjabat HE, Paterson DL (2011) Escherichia coli O25b-ST131: a pandemic, multiresistant, community-associated strain. J Antimicrob Chemother 66:1–14

2. Pitout JDD, Laupland KB (2008) Enterobacteriaceae: an emerging public-health concern. Lancet Infect Dis 8:159–166

3. Peirano G, Pitout JDD (2010) Molecular epidemiology of Escherichia coli producing CTX-M beta-lactamases: the worldwide emergence of clone ST131 O25:H4. Int J Antimicrob Agents 35:316–321

4. Bush K, Jacoby GA (2010) Updated functional classification of beta-lactamases. Antimicrob Agents Chemother 54:969–976

5. Naseer U, Sundsfjord A (2011) The CTX-M conundrum: dissemination of plasmids and Escherichia coli clones. Microb Drug Resist (Larchmont, NY) 17:83–97

6. Pitout JDD (2012) Extraintestinal pathogenic Escherichia coli: an update on antimicrobial resistance, laboratory diagnosis and treatment. Expert Rev Anti-Infect Ther 10:1165–1176

7. ECDC (2012) Antimicrobial resistance surveillance in Europe. In: Annual report of the European Antimicrobial Resistance Surveillance Network (EARS-Net) 2012. https://ecdc.europa.eu/sites/portal/files/media/en/publications/Publications/antimicrobial-resistance-surveillance-europe-2012.pdf

8. Bush K (2010) Alarming β-lactamase-mediated resistance in multidrug-resistant Enterobacteriaceae. Curr Opin Microbiol 13:558–564

9. Pitout JDD (2012) Extraintestinal pathogenic Escherichia coli: a combination of virulence with antibiotic resistance. Front Microbiol 3:9

10. Falagas ME, Karageorgopoulos DE (2009) Extended-spectrum beta-lactamase-producing organisms. J Hosp Infect 73:345–354

11. Dantas G, Sommer MO, Degnan PH et al (2013) Experimental approaches for defining functional roles of microbes in the human gut. Annu Rev Microbiol 67:157–176

12. Grazul H, Kanda LL, Gondek D (2016) Impact of probiotic supplements on microbiome diversity following antibiotic treatment of mice. Gut Microbes 7:101–114

13. Donskey CJ (2006) Antibiotic regimens and intestinal colonization with antibiotic-resistant gram-negative bacilli. Clin Infect Dis 43(Suppl 2):S62–S69

14. Leatham MP, Banerjee S, Autieri SM et al (2009) Precolonized human commensal Escherichia coli strains serve as a barrier to E. coli O157:H7 growth in the streptomycin-treated mouse intestine. Infect Immun 77:2876–2886

15. Nielsen KL, Dynesen P, Larsen P et al (2014) Faecal Escherichia coli from patients with E. coli urinary tract infection and healthy controls who have never had a urinary tract infection. J Med Microbiol 63:582–589

16. Raum E, Lietzau S, von Baum H et al (2008) Changes in Escherichia coli resistance patterns during and after antibiotic therapy: a longitudinal study among outpatients in Germany. Clin Microbiol Infect 14:41–48

17. Faure S, Perrin-Guyomard A, Delmas JM et al (2010) Transfer of plasmid-mediated CTX-M-9 from Salmonella enterica serotype Virchow to Enterobacteriaceae in human flora-associated rats treated with cefixime. Antimicrob Agents Chemother 54:164–169

18. Feld L, Schjørring S, Hammer K et al (2008) Selective pressure affects transfer and establishment of a Lactobacillus plantarum resistance plasmid in the gastrointestinal environment. J Antimicrob Chemother 61:845–852

19. Bailey JK, Pinyon JL, Anantham S et al (2010) Commensal Escherichia coli of healthy humans:

a reservoir for antibiotic-resistance determinants. J Med Microbiol 59:1331–1339

20. Kristiansson C, Grape M, Gotuzzo E et al (2009) Socioeconomic factors and antibiotic use in relation to antimicrobial resistance in the Amazonian area of Peru. Scand J Infect Dis 41:303–312

21. Knudsen JD, Frimodt-møller N (2001) Animal models in bacteriology, vol 9. Karger, Basel, pp 1–14

22. Jakobsen L, Cattoir V, Jensen KS et al (2012) Impact of low-level fluoroquinolone resistance genes qnrA1, qnrB19 and qnrS1 on ciprofloxacin treatment of isogenic Escherichia coli strains in a murine urinary tract infection model. J Antimicrob Chemother 67:2438–2444

23. Krych L, Hansen CHF, Hansen AK et al (2013) Quantitatively different, yet qualitatively alike: a meta-analysis of the mouse core gut microbiome with a view towards the human gut microbiome. PLoS One 8:e62578

24. Leatham MP, Stevenson SJ, Gauger EJ et al (2005) Mouse intestine selects nonmotile flhDC mutants of Escherichia coli MG1655 with increased colonizing ability and better utilization of carbon sources. Infect Immun 73:8039–8049

25. Schjørring S, Struve C, Krogfelt KA (2008) Transfer of antimicrobial resistance plasmids from Klebsiella pneumoniae to Escherichia coli in the mouse intestine. J Antimicrob Chemother 62:1086–1093

26. Hoyen CK, Pultz NJ, Paterson DL et al (2003) Effect of parenteral antibiotic administration on establishment of intestinal colonization in mice by Klebsiella pneumoniae strains producing extended-spectrum β-lactamases. Antimicrob Agents Chemother 47:3610–3612

27. Pultz MJ, Donskey CJ (2007) Effects of imipenem-cilastatin, ertapenem, piperacillin-tazobactam, and ceftriaxone treatments on persistence of intestinal colonization by extended-spectrum-beta-lactamase-producing Klebsiella pneumoniae strains in mice. Antimicrob Agents Chemother 51:3044–3045

28. Donskey CJ, Helfand MS, Pultz NJ et al (2004) Effect of parenteral fluoroquinolone administration on persistence of vancomycin-resistant Enterococcus faecium in the mouse gastrointestinal tract. Antimicrob Agents Chemother 48:326–328

29. Boetius Hertz F, Lobner-Olesen A, Frimodt-Moller N (2014) Antibiotic selection of Escherichia coli sequence type 131 in a mouse intestinal colonization model. Antimicrob Agents Chemother 58:6139–6144

30. Møller AK, Leatham MP, Conway T et al (2003) An Escherichia coli MG1655 lipopolysaccharide deep-rough core mutant grows and survives in mouse cecal mucus but fails to colonize the mouse large intestine. Infect Immun 71:2142–2152

31. Perez F, Pultz MJ, Endimiani A et al (2011) Effect of antibiotic treatment on establishment and elimination of intestinal colonization by KPC-producing Klebsiella pneumoniae in mice. Antimicrob Agents Chemother 55:2585–2589

32. Stiefel U, Pultz NJ, Donskey CJ (2007) Effect of carbapenem administration on establishment of intestinal colonization by vancomycin-resistant enterococci and Klebsiella pneumoniae in mice. Antimicrob Agents Chemother 51:372–375

33. Dobrindt U, Hacker JH, Svanborg C (2013) Between pathogenicity and commensalism, Current topics in microbiology and immunology. Springer, New York

34. Oteo J, Navarro C, Cercenado E et al (2006) Spread of Escherichia coli strains with high-level cefotaxime and ceftazidime resistance between the community, long-term care facilities, and hospital institutions. J Clin Microbiol 44:2359–2366

35. Tian SF, Chen BY, Chu YZ et al (2008) Prevalence of rectal carriage of extended-spectrum b-lactamase-producing Escherichia coli among elderly people in community settings in China. Can J Microbiol 54:781–785

36. Wickramasinghe NH, Xu L, Eustace A et al (2012) High community faecal carriage rates of CTX-M ESBL-producing Escherichia coli in a specific population group in Birmingham, UK. J Antimicrob Chemother 67:1108–1113

37. Luvsansharav U-O, Hirai I, Nakata A et al (2012) Prevalence of and risk factors associated with faecal carriage of CTX-M β-lactamase-producing Enterobacteriaceae in rural Thai communities. J Antimicrob Chemother 67:1769–1774

38. Nicolas-Chanoine M-H, Gruson C, Bialek-Davenet S et al (2013) 10-Fold increase (2006–11) in the rate of healthy subjects with extended-spectrum β-lactamase-producing Escherichia coli faecal carriage in a Parisian check-up centre. J Antimicrob Chemother 68:562–568

39. Nielsen KL, Godfrey PA, Stegger M et al (2014) Selection of unique Escherichia coli clones by random amplified polymorphic DNA (RAPD): evaluation by whole genome sequencing. J Microbiol Methods 103:101–103

40. Zak O, Sande MA, O'Reilly T (1999) Handbook of animal models of infection. Academic, San Diego

41. Erlendsdottir H, Knudsen JD, Odenholt I et al (2001) Penicillin pharmacodynamics in four experimental pneumococcal infection models. Antimicrob Agents Chemother 45:1078–1085

42. Knudsen JD, Fuursted K, Frimodt-Møller N et al (1997) Comparison of the effect of cefepime with four cephalosporins against pneumococci with various susceptibilities to penicillin, in vitro and in the mouse peritonitis model. J Antimicrob Chemother 40:679–686

43. Sandberg A, Jensen KS, Baudoux P et al (2010) Intra- and extracellular activities of dicloxacillin against Staphylococcus aureus in vivo and in vitro. Antimicrob Agents Chemother 54:2391–2400

44. Kerrn MB, Frimodt-Møller N, Espersen F (2003) Effects of sulfamethizole and amdinocillin against Escherichia coli strains (with various susceptibilities) in an ascending urinary tract infection mouse model. Antimicrob Agents Chemother 47:1002–1009

45. Asahi Y, Miyazaki S, Yamaguchi K (1995) In vitro and in vivo antibacterial activities of BO-2727, a new carbapenem. Antimicrob Agents Chemother 39:1030–1037

46. Imhof A (1996) Continuous monitoring of oxygen concentrations in several systems for cultivation of anaerobic bacteria. J Clin Microbiol 34:1646–1648

47. Andrews JM (2001) Determination of minimum inhibitory concentrations. J Antimicrob Chemother 48(Suppl 1):5–16

Chapter 11

Transcriptional Profiling *Mycobacterium tuberculosis* from Patient Sputa

Leticia Muraro Wildner, Katherine A. Gould, and Simon J. Waddell

Abstract

The emergence of drug resistance threatens to destroy tuberculosis control programs worldwide, with resistance to all first-line drugs and most second-line drugs detected. Drug tolerance (or phenotypic drug resistance) is also likely to be clinically relevant over the 6-month long standard treatment for drug-sensitive tuberculosis. Transcriptional profiling the response of *Mycobacterium tuberculosis* to antimicrobial drugs offers a novel interpretation of drug efficacy and mycobacterial drug-susceptibility that likely varies in dynamic microenvironments, such as the lung. This chapter describes the noninvasive sampling of tuberculous sputa and techniques for mRNA profiling *M. tb* bacilli during patient therapy to characterize real-world drug actions.

Key words *Mycobacterium tuberculosis*, Mycobacteria, Transcriptional profiling, Transcriptome, Sputum, RNA extraction, RNA amplification, Microarray analysis

1 Introduction

Transcriptional profiling is an approach that can assist in understanding how cells respond to their changing environment. Gene expression profiling has been applied to *Mycobacterium tuberculosis* to define adaptations to: antimicrobial drug exposure in vitro [1, 2]; the changing macrophage intracellular environment [3–5]; and animal models of disease [6]. Genome-wide mRNA patterns have also captured snapshots of human host–pathogen interplay from expectorated sputa [7] or lung resection tissue [8]. More recently, transcriptional profiling bacilli from sputa has allowed *M. tb* responses to standard regimen drug therapy to be mapped in a clinical setting, revealing insights into the physiological state of *M. tb* expectorated from the lungs and understanding drug efficacy in patients [9, 10]. Multiple techniques exist for mapping mRNA on a genome-wide or near genome-wide scale from quantitative RT-PCR panels and multiplex detection methodologies, to microarrays, to RNAseq using the range of next-generation

Stephen H. Gillespie (ed.), *Antibiotic Resistance Protocols*, Methods in Molecular Biology, vol. 1736,
https://doi.org/10.1007/978-1-4939-7638-6_11, © Springer Science+Business Media, LLC 2018

sequencing platforms. Microarray hybridization as an established technology continues to be useful in some settings. For example, the sequence specificity of target-probe hybridization allows a genome-wide profile to be generated from samples against a background of other RNAs, such as *M. tb* bacilli recovered from human sputum.

This chapter describes the isolation and purification of mycobacterial RNA from human expectorated sputa using a differential lysis technique, followed by RNA amplification using a modified Eberwine in vitro transcription method. The amplified RNA is then chemically labeled with fluorophore using nonenzymatic technology and hybridized to microarrays, designed by the Bacterial Microarray Group at St George's and manufactured by Agilent Technologies.

2   Materials

2.1 Mycobacterial RNA Extraction and Amplification

1. 5 M GTC solution: 5 M guanidine thiocyanate, 0.5% w/v sodium *N*-lauryl sarcosine, 25 mM sodium citrate pH 7, 1% v/v Tween 80, 0.1 M β-mercaptoethanol (*see* **Note 1**).

2. 30 mL V-bottom universal tubes (*see* **Note 2**).

3. Benchtop centrifuge.

4. TRIzol® reagent (Thermo Fisher Scientific) (*see* **Note 3**).

5. 2 mL screw-capped tubes with O-rings containing 0.1 mm silica beads (Lysing matrix B, MP Biomedicals).

6. Reciprocal shaker (FastPrep-24, MP Biomedicals) or equivalent.

7. 1.5 mL nuclease-free tubes.

8. Microcentrifuge.

9. Chloroform (molecular grade).

10. Isopropanol (molecular grade).

11. Ethanol (molecular grade) 100% and 70%.

12. RNase-free water.

13. RNeasy® mini columns (Qiagen).

14. RNase-free DNase kit (Qiagen).

15. NanoDrop Spectrophotometer (Thermo Fisher Scientific).

16. Agilent 2100 Bioanalyzer system (Agilent Technologies).

17. MessageAmp™ II-Bacteria RNA amplification kit (Thermo Fisher Scientific).

18. PCR thermal cycler, heat block or incubator.

19. Vortex mixer.

2.2 Sample Labeling and Microarray Hybridization

1. Amber-colored tubes, 0.5 and 1.5 mL (Alpha Laboratories).

2. Nuclease-free water.

3. Kreatech Universal Linkage System (ULS™) Fluorescent Labeling Kit for Agilent microarrays with Cy3 and KREApure columns (Leica Biosystems).

4. PCR thermal cycler.

5. Heat block.

6. Oven set at 37 °C.

7. Microcentrifuge.

8. Vortex mixer.

9. RNA Fragmentation Reagents (Thermo Fisher Scientific).

10. Hi-RPM Hybridization Buffer (2×) (Agilent Technologies).

11. Gene Expression Wash Kit (Gene Expression Wash Buffer 1, Gene Expression Wash Buffer 2, 10% Triton X-102) (Agilent Technologies).

12. Hybridization gasket slide kit (8 microarrays per slide format) (Agilent Technologies).

13. Microarray slides (Agilent eArray 60-mer SurePrint HD format, Agilent Technologies) in this case a 8x15k *M. tb* complex pan-genome microarray generated by the Bacterial Microarray Group at St. George's (ArrayExpress accession number ABUGS-41) [10, 11].

14. Microarray hybridization chamber (Agilent Technologies).

15. Microarray hybridization oven with rotator rack (Agilent Technologies).

16. Glass slide-staining trough (×3) with swing handle slide rack (×1).

17. Magnetic stir plate and magnetic stir bar (×2).

18. Ozone-barrier slide cover kit (Agilent Technologies).

19. Agilent Microarray Scanner; G4900DA, G2565CA or G2565BA (Agilent Technologies) with Agilent Feature Extraction Software.

3 Methods

3.1 Mycobacterial RNA Extraction and Amplification

1. Immediately after expectoration, add the patient sputa to 4 volumes of 5 M GTC solution and mix. Aliquot the sputa/GTC mixture into 30 mL universal tubes and spin at 1800 × *g* in a benchtop centrifuge for 30 min (*see* **Notes 4** and **5**).

2. Remove the supernatant. Combine sample pellets (if using multiple tubes per sample) in approximately 15 mL GTC solu-

tion, washing the universals with GTC solution to ensure all bacilli are recovered. Spin in a single universal tube per sample at $1800 \times g$ in a benchtop centrifuge for 20 min and remove the supernatant.

3. Resuspend the pellet in 1 mL TRIzol and transfer the suspension to a 2 mL screw-capped tube containing 0.5 mL of 0.1 mm silica beads. Wash the universal tube with an additional 200 µL TRIzol to recover all bacilli in 1.2 mL final volume (*see* **Notes 6** and **7**).

4. Lyse the bacteria using a reciprocal shaker at speed 6.5 for 45 s, then incubate at room temperature for 10 min.

5. Add 200 µL chloroform to each sample, vortex for 30 s, then incubate at room temperature for 10 min to partition the aqueous and phenolic phases. Centrifuge at $15{,}000 \times g$ in a microcentrifuge for 15 min at 4 °C.

6. Transfer the aqueous phase to a new 1.5 mL tube, add an equal volume of chloroform and centrifuge at $15{,}000 \times g$ in a microcentrifuge for 15 min at 4 °C.

7. Transfer the aqueous phase to a fresh nuclease-free 1.5 mL tube, add 0.8 volume of isopropanol and mix by inverting. Incubate overnight at −20 °C to precipitate the nucleic acids (*see* **Note 8**).

8. Centrifuge the samples in a microcentrifuge at $15{,}000 \times g$ for 20 min at 4 °C to pellet the nucleic acid (*see* **Note 9**).

9. Remove the supernatant carefully by pipetting and wash the pellet with 500 µL cold 70% ethanol. Centrifuge in a microcentrifuge at $15{,}000 \times g$ for 15 min at 4 °C.

10. Remove the ethanol by pipetting, respin the tubes briefly and remove any additional liquid.

11. Air-dry the pellet for 5–10 min at room temperature and resuspend in 100 µL RNase-free water. Store briefly on ice and proceed to RNA cleanup using RNeasy Mini Columns, with buffers prepared and stored according to manufacturer's instructions.

12. Add 350 µL RNeasy RLT buffer to each 100 µL sample and mix thoroughly by pipetting.

13. Add 250 µL 100% ethanol, mix thoroughly by pipetting and apply immediately to an RNeasy Mini Column placed in a 2 mL collection tube. Centrifuge in a microcentrifuge at $9000 \times g$ for 15 s, discard the flow-through.

14. Apply 350 µL RNeasy RW1 buffer to the column, centrifuge in a microcentrifuge at $9000 \times g$ for 15 s and discard the flow-through.

15. DNase I treat the samples to remove contaminating DNA, using the Qiagen RNase-free DNase kit. Apply 80 μL of DNase I: buffer RDD mix (10 μL DNase I, 70 μL RDD buffer, prepared immediately before use) directly onto the column matrix. Incubate at room temperature for 15 min.

16. Wash the column with 350 μL RW1 buffer and centrifuge in a microcentrifuge at 9000 × g for 15 s. Discard the flow-through.

17. Wash the column with 500 μL RPE buffer, centrifuge in a microcentrifuge at 9000 × g for 15 s. Discard the flow-through.

18. Pipette 500 μL RPE buffer onto the column matrix and centrifuge at 9000 × g in a microcentrifuge for 2 min. Place the column into a fresh 2 mL collection tube and centrifuge in a microcentrifuge for an additional 1 min at 15,000 × g to prevent carry-over of wash buffer.

19. Transfer the column to a fresh nuclease-free 1.5 mL tube and add 30 μL RNase-free water directly onto the column matrix. Incubate at room temperature for 2 min and centrifuge at 9000 × g in a microcentrifuge for 1 min to elute the RNA. Reapply the eluate to the column, incubate for further 2 min and centrifuge at 9000 × g for 1 min (*see* **Note 10**).

20. Quantify the RNA samples and assess the integrity of the RNA using the NanoDrop Spectrophotometer and Agilent 2100 Bioanalyzer (or similar) following manufacturers' instructions.

21. Store the samples at −70 °C (*see* **Note 11**) or continue with amplification of the RNA using the Bacteria MessageAmp II system.

22. Adjust RNA sample volume to 5 μL with nuclease-free water (*see* **Note 12**). Incubate for 10 min at 70 °C, before placing on ice for 3 min. Briefly centrifuge, then add 5 μL polyadenylation master mix (1 μL 10× buffer, 1 μL RNase inhibitor, 0.5 μL ATP, 1 μL PAP, 1.5 μL nuclease-free water) and incubate at 37 °C for 15 min. Place on ice before proceeding immediately to the next step.

23. Add 10 μL reverse transcription master mix (1 μL 10× first strand buffer, 1 μL T7 oligo-dT, 4 μL dNTP mix, 1 μL ArrayScript reverse transcriptase, 3 μL nuclease-free water), mix gently by pipetting and incubate at 42 °C for 2 h. Place the reactions on ice, then centrifuge briefly.

24. Add 80 μL second strand master mix (10 μL 10× second strand buffer, 4 μL dNTP mix, 1 μL RNase H, 2 μL DNA polymerase, 63 μL nuclease-free water), mix by pipetting then incubate at 16 °C for 2 h. Return to ice, centrifuge briefly.

25. To purify the cDNA, add 250 µL cDNA binding buffer to each sample, and mix by pipetting. Transfer the samples onto the cDNA filter cartridge matrix and centrifuge at 9000 × *g* for 1 min in a microcentrifuge. Discard the flow-through. Wash with 500 µL wash buffer and centrifuge at 9000 × *g* for 1 min. Discard the flow-through and then centrifuge the columns for an additional minute to remove excess wash buffer. Transfer the filter cartridges into clean cDNA elution tubes, and elute by adding 18 µL preheated 55 °C nuclease-free water to the column matrix. Incubate at room temperature for 2 min then centrifuge at 9000 × *g* for 1.5 min.

26. Add 24 µL unmodified In Vitro Transcription (IVT) master mix (4 µL 10× reaction buffer, 4 µL T7 ATP, 4 µL T7 CTP, 4 µL T7 GTP, 4 µL T7 UTP, 4 µL T7 enzyme) to each sample (total reaction volume 40 µL), mix gently, and incubate at 37 °C for 16 h (*see* **Note 13**). After incubation, make up the sample volume to 100 µL by adding 60 µL nuclease-free water. Place on ice.

27. To purify the amplified RNA (aRNA), add 350 µL aRNA binding buffer to each sample and mix by pipetting. Add 250 µL 100% ethanol and mix by gently pipetting. Transfer onto the aRNA filter cartridge matrix and centrifuge at 9000 × *g* in a microcentrifuge for 1 min. Discard the flow-through.

28. Wash the columns with 650 µL Wash buffer, before centrifuging at 9000 × *g* for 1 min. Discard the flow-through and recentrifuge the columns at 9000 × *g* for an additional minute to remove excess buffer.

29. Transfer the filter cartridges into fresh aRNA elution tubes. Elute the aRNA by adding 50 µL preheated 55 °C nuclease-free water, incubate at room temperature for 2 min and centrifuge for 1.5 min at 9000 × *g*. Repeat elution a second time with a further 50 µL nuclease-free water. Estimate aRNA yield using the NanoDrop spectrophotometer, and store aRNA at −70 °C.

3.2 Sample Labeling and Microarray Hybridization

1. Sample labeling using the nonenzymatic Kreatech Universal Linkage System (ULS). For each sample, add 1 µg aRNA, 1 µL ULS-Cy3, 1.5 µL 10× Labelling solution and adjust volume to 15 µL. Mix by pipetting and incubate at 85 °C for 15 min (*see* **Notes 14** and **15**).

2. Transfer the samples to ice and incubate for 3 min. Centrifuge briefly.

3. Remove nonreacted ULS-Cy3 using KREApure columns. Resuspend the KREApure column material by briefly mixing using a vortex mixer.

4. Loosen cap ¼ turn, snap off the bottom closure and place the column into a 2 mL collection tube.

5. Centrifuge the column at $15{,}000 \times g$ in a microcentrifuge for 2 min, and discard the flow-through and cap. Place the column back into the same collection tube.

6. Add 300 µL nuclease-free water to the column and spin for 2 min at $15{,}000 \times g$. Discard the flow-through and collection tube, and transfer the column to a new 1.5 mL tube.

7. Add labeled aRNA to the column matrix and centrifuge for 2 min at $15{,}000 \times g$ in a microcentrifuge. Optional, use 1.5 µL of each sample eluate to measure Cy3-incorporation using the NanoDrop Spectrophotometer.

8. Transfer the aRNA to a 0.5 mL nuclease-free tube and fragment by adding 1.5 µL 10× fragmentation buffer. Incubate at 70 °C for 15 min.

9. Centrifuge briefly and add 1.5 µL stop solution, mix by pipetting and place on ice.

10. Briefly centrifuge. Prepare the hybridization solution adding 11.3 µL labeled aRNA, 11.2 µL KREAblock blocking agent and 22.5 µL Agilent 2× Hybridization buffer to a fresh 0.5 mL tube. Mix thoroughly (by vortexing) being careful not to introduce bubbles (*see* **Note 16**).

11. Place a clean gasket slide (to match microarray layout, in this instance 8×15k) into the hybridization chamber base (*see* **Note 17**).

12. Slowly dispense 40 µL hybridization solution onto the gasket well in a "drag and dispense" manner. Do not allow the liquid to touch the edges of the gasket well and try not to introduce bubbles while pipetting. Load the rest of the samples into the remaining gasket wells (*see* **Notes 18** and **19**).

13. Place the active side of the microarray slide face down onto the gasket slide (numeric barcode facing up, Agilent-labeled barcode facing down) (*see* **Note 20**).

14. Add the hybridization chamber cover, slide the clamp into place and hand tighten.

15. Rotate the assembled hybridization chamber to check that the air bubble in each well of the gasket moves the sample across the microarray surface. Tap to move stationary air bubbles if necessary (*see* **Note 21**).

16. Place the hybridization chamber into the rotator rack of the hybridization oven set to 65 °C. Rotate at 20 rpm and incubate overnight (17 h). Place 400 mL Gene Expression Wash buffer 2 in a sealed bottle and incubate at 37 °C overnight along with an empty staining trough.

17. After hybridization, fill one staining trough with ~400 mL room temperature Agilent Gene Expression Wash buffer 1 (trough 1) and fill a second with Agilent Gene Expression Wash buffer 1 to cover a slide rack (trough 2). Add a slide rack and a stir bar to trough 2.

18. Remove the slide-gasket sandwich from the hybridization chamber base and submerge in trough 1 without letting go of the slides. Using tweezers, pry the sandwich open from the barcode end keeping the slide numeric barcode facing up. Let the gasket slide drop to the bottom of the staining trough whilst keeping hold of the microarray slide (*see* **Note 22**).

19. Transfer the microarray slide to the rack in trough 2 (containing Wash buffer 1) and stir using the magnetic plate for 1 min.

20. Before Wash 1 is complete, fill the preheated staining trough (trough 3) with the preheated Agilent Gene Expression Wash buffer 2 and add a stir bar.

21. Transfer the slide rack from trough 2 to trough 3 and stir for 1 min at 37 °C (*see* **Note 23**).

22. Slowly remove the slide rack from Wash buffer 2 minimizing droplets forming on the slides. Rest the slide rack on a paper towel.

23. Transfer slides immediately to Agilent slide holders and add ozone-barrier covers (if using Scanners G2565CA or G2565BA). Scan the slides immediately using Agilent Microarray Scanner (G4900DA, G2565CA or G2565BA) at 5 μm resolution. Extract data from image files using Agilent Feature Extraction software.

4 Notes

1. To prepare 500 mL of 5 M GTC solution, add guanidine thiocyanate powder to a graduated 500 mL flask. Add approximately 200 mL distilled water, mix and leave in warm room overnight (reaction is endothermic). When GTC powder has dissolved, add remaining constituents, except β-mercaptoethanol. Adjust volume to 500 mL by adding distilled water. Store at room temperature away from direct sunlight. Add β-mercaptoethanol before use. Discard if GTC solution develops a yellow color.

2. V-bottom universal tubes preferred to Falcon tubes as the bacilli centrifuge into tighter pellets.

3. TRI Reagent® (Sigma-Aldrich) or similar products also acceptable.

4. Plunge sputa into GTC solution within 5 min of sampling to retain a representative RNA profile. Mycobacterial transcrip-

tion ceases on addition of GTC solution and nuclease action is minimized to stabilize the RNA signature. Solutions/centrifugations do not need to be chilled. Mycobacteria should not lyse in the presence of GTC solution however eukaryotic cells and other bacteria may. This serves to reduce background RNA and allows accurate quantification of mycobacteria-derived RNA from sputa; other RNAs will be found in the sputa/GTC supernatant. If the sputa/GTC solution becomes viscous, vortex, syringe, or add additional GTC solution to ensure a pellet is able to form during centrifugation.

5. This RNA extraction methodology may be applied to mycobacterial samples from in vitro axenic or intracellular infection models. If statistical testing is to be applied to the transcriptional dataset, ensure that appropriate comparator conditions and sample replicates are collected using the same RNA extraction methodology.

6. If performing RNA extraction in batches, which is recommended to ensure consistency, add 1 mL TRIzol to each pellet, transfer to a 2 mL screw-capped tube and store at −70 °C. Defrost in batches to resume RNA extraction and RNA amplification.

7. TRIzol effectively sterilizes pathogenic mycobacteria, so the rest of this protocol may be conducted outside Category Three Containment conditions—this should be validated according to local biosafety guidelines.

8. It is not necessary to add additional salt to increase precipitation efficiency. For some applications, such as isolating small RNAs, skip this precipitation step and proceed directly to purification using sRNA-compatible columns.

9. A white nucleic acid pellet may be visible but this is not always the case.

10. Elution volumes should be a minimum of 30 μL and a maximum of 100 μL. A second elution is recommended to increase RNA yield.

11. Assess quantity and quality of the RNA immediately (before freezing) or, to avoid freeze–thaw cycles, save 2 μL aliquots of each RNA preparation for NanoDrop and Bioanalyzer analysis at a later date.

12. Total RNA input may range from 5 to 500 ng. The input RNA for all samples should be equal. Amplification changes the RNA profile, so amplified RNA should never be compared directly to unamplified RNA [12]. All samples to be compared should be amplified together to avoid introducing unnecessary technical variation.

13. Use an incubator for the IVT reaction or a PCR-block with variable heated-lid, so condensation does not build up on the tube lids overnight.

14. Use a ratio of 1 μL ULS-Cy3 per 1 μg aRNA. Prepare the correct number of samples per microarray slide. In this example, an Agilent Technologies SurePrint HD 8×15k slide, so label samples in batches of 8.

15. The design of the microarray should be taken into consideration when choosing a technique to incorporate Cy3 before hybridization. In this protocol the ULS labeling system directly labels amplified RNA to hybridize to an 8×15k Agilent Technologies *M. tb* complex microarray slide (ArrayExpress accession number ABUGS-41). To hybridize unamplified RNA to the same array would require conversion to cDNA incorporating Cy3-dCTP [13].

16. The 2× Hybridization buffer contains surfactant that easily forms bubbles, so vortex mix carefully.

17. To avoid damaging the microarray, maintain a clean work area and handle the slides carefully by the edges, never touching the surfaces. Always wear powder-free gloves.

18. The hybridization solution is applied onto the gasket slide rather than directly onto the microarray slide, which will be placed onto the gasket slide and samples. The surface tension of the liquid allows the sample to be pipetted into the centre of each well of the gasket without touching the sides. When the microarray is lowered on top, an air bubble forms around the inside edge of the gasket, which serves to mix the sample during hybridization.

19. Gasket slides come in four different formats: 1, 2, 4, and 8×. The hybridization volumes detailed in this protocol are for use with 8× gasket slides. If using 1, 2, or 4× format, apply 490, 245, or 100 μL of the hybridization solution respectively.

20. Line up the slide between finger and thumb a few millimeters above and parallel to the gasket, drop into place. The samples should be sandwiched between the gasket and microarray slide. Tap the top of the microarray slide with a pipette tip to ensure slide contact with all the samples. There should be an air pocket surrounding each sample volume within each well of the gasket.

21. Stationary air bubbles will compromise the uniformity of the array hybridization and may lead to loss of data.

22. Ensure that the array-gasket sandwich stays completely submerged in the wash buffer during disassembly.

23. Wash buffer 2 is a higher stringent buffer than Wash 1, therefore Wash 2 is time sensitive and careful timekeeping is important.

Acknowledgments

L.M.W. was funded by a Brazilian government agency CAPES (Coordination for the Improvement of Higher Education Personnel) PhD visiting fellowship [99999.005648/2014-09]. K.A.G. acknowledges funding from the Wellcome Trust for the Bacterial Microarray Group at St. George's [062511, 080039, and 086547]. S.J.W. was supported by the Wellcome Trust [204538/Z/16/Z] and the PreDiCT-TB consortium (http://www.predict-tb.eu) which is funded from the Innovative Medicines Initiative Joint Undertaking under grant agreement No 115337, resources of which are composed of financial contribution from the European Union's Seventh Framework Programme (FP7/2007-2013) and EFPIA companies' in-kind contribution.

References

1. Boshoff HI, Myers TG, Copp BR, McNeil MR, Wilson MA, Barry CE III (2004) The transcriptional responses of *Mycobacterium tuberculosis* to inhibitors of metabolism: novel insights into drug mechanisms of action. J Biol Chem 279(38):40174–40184

2. Waddell SJ, Stabler RA, Laing K, Kremer L, Reynolds RC, Besra GS (2004) The use of microarray analysis to determine the gene expression profiles of *Mycobacterium tuberculosis* in response to anti-bacterial compounds. Tuberculosis (Edinb) 84(3–4):263–274

3. Schnappinger D, Ehrt S, Voskuil MI, Liu Y, Mangan JA, Monahan IM, Dolganov G, Efron B, Butcher PD, Nathan C, Schoolnik GK (2003) Transcriptional adaptation of *Mycobacterium tuberculosis* within macrophages: insights into the phagosomal environment. J Exp Med 198(5):693–704

4. Rohde KH, Veiga DF, Caldwell S, Balazsi G, Russell DG (2012) Linking the transcriptional profiles and the physiological states of *Mycobacterium tuberculosis* during an extended intracellular infection. PLoS Pathog 8(6):e1002769. https://doi.org/10.1371/journal.ppat.1002769. PPATHOGENS-D-11-02225 [pii]

5. Tailleux L, Waddell SJ, Pelizzola M, Mortellaro A, Withers M, Tanne A, Castagnoli PR, Gicquel B, Stoker NG, Butcher PD, Foti M, Neyrolles O (2008) Probing host pathogen cross-talk by transcriptional profiling of both *Mycobacterium tuberculosis* and infected human dendritic cells and macrophages. PLoS One 3(1):e1403

6. Talaat AM, Ward SK, Wu CW, Rondon E, Tavano C, Bannantine JP, Lyons R, Johnston SA (2007) Mycobacterial bacilli are metabolically active during chronic tuberculosis in murine lungs: insights from genome-wide transcriptional profiling. J Bacteriol 189(11):4265–4274. https://doi.org/10.1128/JB.00011-07. JB.00011-07 [pii]

7. Garton NJ, Waddell SJ, Sherratt AL, Lee SM, Smith RJ, Senner C, Hinds J, Rajakumar K, Adegbola RA, Besra GS, Butcher PD, Barer MR (2008) Cytological and transcript analyses reveal fat and lazy persister-like bacilli in tuberculous sputum. PLoS Med 5(4):e75

8. Rachman H, Strong M, Ulrichs T, Grode L, Schuchhardt J, Mollenkopf H, Kosmiadi GA, Eisenberg D, Kaufmann SH (2006) Unique transcriptome signature of *Mycobacterium tuberculosis* in pulmonary tuberculosis. Infect Immun 74(2):1233–1242

9. Walter ND, Dolganov GM, Garcia BJ, Worodria W, Andama A, Musisi E, Ayakaka I, Van TT, Voskuil MI, de Jong BC, Davidson RM, Fingerlin TE, Kechris K, Palmer C, Nahid P, Daley CL, Geraci M, Huang L, Cattamanchi A, Strong M, Schoolnik GK, Davis JL (2015) Transcriptional adaptation of drug-tolerant *Mycobacterium tuberculosis* during treatment of human tuberculosis. J Infect Dis. https://doi.org/10.1093/infdis/jiv149. jiv149 [pii]

10. Honeyborne I, McHugh TD, Kuittinen I, Cichonska A, Evangelopoulos D, Ronacher K, van Helden PD, Gillespie SH, Fernandez-Reyes D, Walzl G, Rousu J, Butcher PD, Waddell SJ (2016) Profiling persistent tubercule bacilli from patient sputa during therapy predicts early drug efficacy. BMC Med 14:68. https://doi.org/10.1186/s12916-016-0609-3

11. Chatterjee A, Saranath D, Bhatter P, Mistry N (2013) Global transcriptional profiling of longitudinal clinical isolates of *Mycobacterium tuberculosis* exhibiting rapid accumulation of drug resistance. PLoS One 8(1):e54717. https://doi.org/10.1371/journal.pone.0054717. PONE-D-12-24527 [pii]

12. Waddell SJ, Laing K, Senner C, Butcher PD (2008) Microarray analysis of defined *Mycobacterium tuberculosis* populations using RNA amplification strategies. BMC Genomics 9(1):94

13. Waddell SJ, Butcher PD (2010) Use of DNA arrays to study transcriptional responses to antimycobacterial compounds. Methods Mol Biol 642:75–91. https://doi.org/10.1007/978-1-60327-279-7_6

Chapter 12

Direct in Gel Genomic Detection of Antibiotic Resistance Genes in S1 Pulsed Field Electrophoresis Gels

Mark A. Toleman

Abstract

S1 pulsed field gel electrophoresis (PFGE) is a method to separate the bacterial chromosome(s) from plasmid nucleic acids. When combined with ethidium bromide staining and UV visualization this method is excellent at assessing the number of plasmids in individual bacterial strains. It is also good at approximating the true size of each individual plasmid when run against a DNA molecular marker. However, downstream applications such as: the location of individual resistance genes on individual plasmids or the chromosome are hampered by very poor transfer of large DNA molecules from agarose gels to adsorbent nylon or nitrocellulose membranes. Herein, we describe a method to directly probe agarose PFGE gels eliminating the necessity of transfer and generating excellent genomic location results.

Key words Direct agarose gel probing, Pulsed field gel electrophoresis, Plasmid detection, Genomic location, Resistance gene mapping

1 Introduction

Pulsed field gel electrophoresis (PFGE) is an electrophoresis method that separates large molecules of DNA in an agarose gel on the basis of their size [1]. The parameters of the electrophoresis module can be altered to separate almost any range of DNA molecules from small plasmids to whole chromosomes. Large pieces of DNA such as conjugative plasmids (above 30 kb) and bacterial chromosomes are typically very fragile and can be sheared by pipetting liquid samples of DNA. For this reason bacterial cultures are lysed and digested in a solid agarose medium, typically as small agarose plugs, which are then inserted into an agarose gel. S1 is a nuclease that primarily has single stranded DNA endonuclease activity but can also degrade double-stranded DNA at high concentration [2]. Plasmids typically exist in several different structural conformations mostly based on the degree of supercoiling of their DNA, many of which do not run at their true molecular size in agarose gels. Partial S1 digestion nicks plasmid DNA at regions

Stephen H. Gillespie (ed.), *Antibiotic Resistance Protocols*, Methods in Molecular Biology, vol. 1736,
https://doi.org/10.1007/978-1-4939-7638-6_12, © Springer Science+Business Media, LLC 2018

that are highly supercoiled where conformational stresses cause small sections of single stranded DNA to be exposed. This removes all supercoiling and leaves a single linear form of the plasmid that runs true to size [2]. The concentration of many different forms of the plasmid into a single linear entity enables visualization of the DNA using ethidium bromide staining when illuminated by UV light. Plasmid sizes can then be estimated against a DNA molecular ladder. The presence and position of any gene in the plasmid or chromosome of each strain is then visualized by autoradiography after probing with the respective radiolabeled gene.

2   Materials

All solutions are prepared using autoclaved double distilled water.

2.1   S1 Digestion Components

1. 1/10 TE buffer; 1 mM Tris–HCl, 0.1 mM EDTA, pH 8. Prepare 1× TE buffer: To 10 mL of Tris 1 M, pH 8, add 2 mL of 0.5 M EDTA, pH 8 and then dilute to 1000 mL with water. Dilute 5 mL of 1× TE to 50 mL to produce 1/10× TE buffer.

2. 1× S1 buffer: 30 mM sodium acetate pH 4.6, 1 mM ZnSo4, 5% glycerol. Prepare 10× S1 buffer by adding 12.3 g of sodium acetate and 0.92 g Zinc acetate to 200 mL of water. Add 250 mL of glycerol and adjust pH to 4.6. Then add water to 500 mL. Dilute 5 mL of 10× S1 buffer to 50 mL to produce 1× S1 buffer. Store 10× buffer aliquoted at −20 °C.

2.2   Gel Running Components

1. Gel and gel running buffer 0.5× TBE: 45 mM Tris–HCl, 45 mM boric acid, 1 mM EDTA. Prepare 10× TBE buffer by adding 108 g Tris base, 55 g boric acid and 40 mL 0.5 M EDTA, pH 8 to 700 mL of water. Adjust pH to 8 with concentrated HCl and make up to 1000 mL and autoclave before use. Add 100 mL of 10× TBE to 1900 mL of water to produce 0.5× TBE.

2.3   In Gel Hybridization Components

1. Denaturing solution: 0.5 M NaOH, 1.5 M NaCl. Add 20 g NaOH and 87.66 g NaCl to 1 L water.

2. Neutralizing solution: 0.5 M Tris–HCl, pH 7.5, 1.5 M NaCl. Dissolve 60.5 g Tris base and 87.6 g NaCl in 800 mL of water adjust to pH 7.5 with concentrated HCl.

3. The prehybridization solution consists of (6× SSC, 0.1% (W/V) polyvinylpyrrolidone, 0.1% Ficoll, 0.5% SDS, 150 µg/mL herring testes DNA, and 1 mL UHT full cream milk made up to 20 mL with water.

4. Random priming: Prime-It II Random Primer Labeling Kit, Agilent Technologies, UK.

5. ^{32}P labeled dCTP (EasyTides Deoxycytidine 5'-triphosphate. (alpha ^{32}P), PerkinElmer, London, UK).

6. Unincorporated nucleotides are removed using a Sephadex G-50 gel filtration column (illustra™ Nick™ Columns Sephadex G-50 DNA grade, GE Healthcare Life Sciences, Little Chalfont, Buckinghamshire, UK).

7. Detection film used is: Lumi-Film Chemiluminescent Detection film, (Roche, Mannheim, Germany).

3 Methods

3.1 S1 Pulsed Field Gel Electrophoresis

1. Agarose plugs of bacterial strains are prepared using two different previously published methods: For nonfermentative bacteria such as *Pseudomonas aeruginosa* and *Acinetobacter baumanii* we use a 2 day protocol [3]; For enteric bacteria such as *E. coli* and *Klebsiella pneumoniae* we use the standard operating procedure for pulseNet PFGE of *E. coli* 0157, H7, *Escherichia coli* non-0157 (STEC), Salmonella serotypes, *Shigella sonnei*, and *Shigella flexneri*. Available on the Centres of Disease Control website: CDC.gov using the web-link Disease Control website: CDC.gov [4].

2. Once the plugs are prepared they are digested with S1 enzyme (*see* **Note 1**). Each individual plug is first washed at room temperature for 20 min (*see* **Note 2**) in 1 mL of 1/10 TE buffer, followed by a second incubation at room temperature for 20 min in 1 mL of 1× S1 buffer. Once the second wash is completed the S1 buffer is removed and the S1 enzyme is added. The S1 enzyme is used at a very dilute concentration. To 6 mL of 1× S1 buffer, 0.5 μL (50 U) of S1 (stock enzyme at a concentration of 100 U/μL) is added. Two-hundred microliters of the enzyme solution is then used to digest each plug such that the 6 mL is enough to digest 30 plugs. The digest is incubated at 37 °C for 45 min. Once digested the plug is removed from the digest solution and cut in half such that each plug is enough for duplicate samples/gels.

3. Using the gel forming equipment provided by the supplier of the PFGE system, 0.9% agarose gels are poured by adding 1 g of PFGE grade agarose to 110 mL of 0.5× TBE buffer (45 mM Tris, 45 mM boric acid, 1 mM EDTA) and microwaving at full power for about 2 min (*see* **Note 3**). Gels are left to set for about 20 min before plugs are loaded on the gel.

4. After digestion (*see* **Note 4**) each plug is cut in half, and half a plug of each strain added to the gel(s) (*see* **Note 5**). Gels are then inserted into the PFGE equipment under the gel running buffer 2 L of 0.5× TBE Gel running parameter are 6v/cm; an

included angle of 120° with an initial switch time of 5 s and a final switch time of 45 s at 10 °C for 22 h on a CHEF II PFGE machine (Bio-Rad). These parameters are ideal for separating DNA of sizes of about 15 kb to approximately 1 mega base. Ethidium bromide is added to the gel running buffer (20 μL of 10 mg/mL solution).

5. Plasmids are visualized after gel running on a UV transilluminator and photographed using a UVP geldoc II imaging system (UVP Cambridge, UK). Plasmid number and sizes are determined by comparison with the molecular size marker.

3.2 In-Gel Hybridization

1. Gels are dried by placing overnight in a drying cabinet at 50C between two sheets of blotting paper (*see* **Note 6**). Once the gels are dried they can be stored for extended lengths of time (>1 year) in a cool dry place between the pieces of blotting paper.

2. Once dried, gels are rehydrated by placing in 200 mL of deionized DNase-free water in a flat-bottomed pyrex glass bowl for 5 min (*see* **Notes 7** and **8**). If the gel is physically stuck to one of the pieces of blotting paper, add the gel and blotting paper to the water. After 5 min the gel is easily pulled away from the blotting paper without damage.

3. The distilled water is discarded and 200 mL of denaturing solution is added to denature the DNA in the gel. This is incubated at room temperature for 45 min (*see* **Note 9**).

4. The denaturing solution is then removed and the gel neutralized by addition of 200 mL of neutralizing solution and incubated for 45 min at room temperature (*see* **Note 9**).

5. The neutralizing solution is removed and the gel placed in a hybridization tube. 20 mL of prehybridization solution is added to block the gel prior to probing. The gel is incubated for at least 24 h at 65 °C (*see* **Note 10**).

6. Probes are made by first amplifying a desired gene by PCR using specific primers and a bacterial strain that contains the desired gene (In our example a resistance gene such as $bla_{CTX-M-15}$) (*see* **Note 11**).

7. The probe is prepared by the random priming labeling method using the purified PCR product prepared above as a template (200 ng in 15 μL water **Note 12**) and radio-active P^{32} dCTP as a label. We use a commercially available kit using the standard protocol provided with the kit (*see* **Note 13**).

8. Once the probe has been labeled, unincorporated $dCTP^{32}$ and unlabeled nucleotides are removed using a Sephadex G50 gravity flow gel filtration column: Briefly the labeled probe

(60 μL) is added to the top of the gel filtration column followed by 320 μL of 0.1 M Tris pH 7.5. The filtrate flows through the column by gravity flow and is collected in a 1.5 mL eppendorf tube and is discarded. A new eppendorf tube is then placed under the column and the labeled DNA is eluted with 430 μL of 0.1 M Tris pH 7.5. This results in unincorporated nucleotides being left on the column. The 430 μL of labeled probe is then boiled for 6 min in a screw cap eppendorf tube and then added to the prehybridized gel and left to hybridize overnight at 65 °C (*see* **Note 13**).

9. Once hybridized the probe is discarded and the gel is washed for 1 h at 65 °C with 2× SSC 0.1% SDS (100mls) and then again for 1 h at 65 °C with 0.1× SSC, 0.1% SDS (100 mL) (*see* **Note 14**).

10. Washed gels are finally removed from the hybridization tubes, washed under a warm tap for a couple of minutes and blotted dry with blotting paper. They are then wrapped in cling film and placed against a sheet of film overnight in a film cassette before developing using standard film development and fixer solutions (*see* **Note 15**) (Fig. 1).

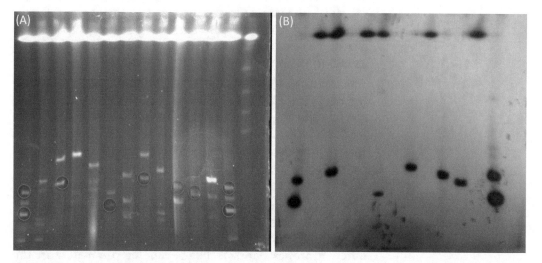

Fig. 1 (**a**) S1 pulsed field gel of 13 *E. coli* strains (lanes 1–13) run against a *Saccharomyces cerevisiae* molecular size standard (Lane 14) and illuminated with ethidium bromide staining under ultraviolet light. The chromosome of each bacterial isolate is the brightest band at the top of the gel and plasmids are visualized as bright bands towards the lower end of the gel. Each *E. coli* isolate has between 1 (lane 11) and 6 (lane 2) plasmids. (**b**) Auto radiograph of (**a**) after probing with a radioactive P^{32}-labeled $bla_{CTX-M-15}$ probe. Note the $bla_{CTXM-15}$ gene is found on the chromosome in *E. coli* isolates (lanes 2, 3, 5, 6, 9, and 12) on multiple plasmids in lanes 1 and 13 and on the chromosome as well as on an individual plasmids in lane 3 and 6. It is only found on the plasmid in lanes 8, 10, and 11. Positive $bla_{CTX-M-15}$ plasmids are highlighted in (**a**) with a blue circle

4   Notes

1. The amount of enzyme used is crucial for the success of the experiment and the activity of this enzyme varies between suppliers. We generally start with 50 U of S1 enzyme in 6 mL of 1× S1 buffer and use 200 μL of this for each whole plug digest. Once the gel is run UV illumination should reveal a bright chromosomal band together with a faint smear of digested chromosomal DNA throughout the length of the gel. If not enough S1 enzyme is used there will be no smear and plasmids are often not visible. If too much S1 is used the whole chromosome is digested leaving a smear of low molecular weight DNA. The ideal amount of S1 should be determined empirically for each new batch of S1 enzyme by using a range of different concentrations of S1 from between 300 U per plug to 10 U per plug and a digest of 45 min at 37 °C.

2. Each standard PFGE gel has 15 wells, which is enough for 14 sample bacterial strains and a molecular standard ladder. Washes and digestion of plugs for one gel or more are conveniently done in a 48 well culture plate.

3. Agarose gel is prepared by microwaving the solution for 2 min and swirling the flask to see if any agarose crystals are not yet dissolved. If agarose crystals are seen the agarose is re microwaved for 10–20 s time spans until no agarose crystals are visualized. The hot agarose is then poured into the mold being careful to remove all bubbles.

4. There is no need to stop the S1 reaction other than removing the plug from the digest as long as the plugs are loaded and run immediately after the digest. Ideally the plugs should be loaded and the electrophoresis run started within 30 min as longer periods will allow the S1 to degrade the chromosomal DNA.

5. The PFGE equipment is designed to run one gel at a time. However, in practice we often run duplicate gels stacked on top of each other. The advantage of this is that two identical gels are prepared such that the same gels can be probed with different probes. Most of the gel-running parameters are the same when one or two gels are run but if two gels are run a little extra running buffer will need to be added to the PFGE tank to ensure that both gels are under running buffer. We also add an additional hour to the run time when running stacked gels.

6. When drying the PFGE gel between pieces of blotting paper, ensure that a weight is placed evenly covering the length and breadth of the gel. This will ensure that the gel does not fold up as it dries.

7. Water used here and throughout method for all buffers is double distilled water which is then autoclaved and kept in 2 L bottles. Failure to autoclave often leads to DNA degradation due to DNases that are often found in double distilled water which is especially found if the water has been stored in direct sunlight.

8. After 5 min of rehydration the rehydrated gel is about 0.5 mm thick and quite tough, being resilient to folding without breaking and is treated from now on the same as you would a nylon membrane.

9. The denaturing and neutralizing solutions can be reused many times without loss of activity. Simply return the solutions back to their bottles after use.

10. Prehybridization is made up from stock solutions of: 5% PVP (0.4 mL), 5% ficoll (0.4 mL), 10 mg/mL herring testes DNA (0.3 mL), 10% SDS (1 mL), 20× SSC (6 mL), and 1 mL of full cream UHT milk. The herring testes DNA is sheared by repetitive pipetting and then added to the prehyb mix. The UHT milk can be purchased from any supermarket. Prehybing can be left from 24 to a maximum of 72 h.

11. When amplifying the DNA to be used as a probe it is a good idea to amplify the desired gene from an organism that is different to the one being probed (e.g., if probing a gel of DNA's derived from *E. coli* strains use a *Klebsiella pneumoniae* strain containing the desired gene as a template for the probe PCR as this will minimize any background hybridization problems).

12. The kit utilizes random primers, which are 9 bp single stranded oligomers. These will bind randomly by chance to about every 200–300 bp of DNA. The minimum sized PCR product for efficient labeling is therefore about 500 bp. However, in practice we find that PCR products between 800 and 1000 bp produce ideal probes.

13. The probe is added to the prehybridization solution that the gel has been prehybridizing in. There is no necessity to change this solution before adding the boiled probe.

14. Washing can be repeated several times to ensure low backgrounds. The gel can even be left washing over a weekend without any discernable reduction in probe signal. During washing the gel can be removed from the hybridization tube and background levels tested by holding a Geiger tube against parts of the gel, which contain no sample (i.e., the small gel section above the loading wells). In this way low backgrounds can be assured before autoradiography.

15. We use trays containing developer and fixer rather than a machine as the film can be left until all bands are visible before fixing.

References

1. Schwartz DC, Cantor CR (1984) Separation of yeast chromosome-sized DNAs by pulsed field gradient gel electrophoresis. Cell 37(1):67–75

2. Barton BM, Harding GP, Zuccarelli AJ (1995) A general method for detecting and sizing large plasmids. Anal Biochem 226:235–240

3. Patzer JA, Walsh TR, Weeks J, Dzierzanowska D, Toleman MA (2009) Emergence and persistence of integrin structures harbouring VIM genes in the Childrens's Memoial Health Institute, Warsaw, Poland, 1998-2006. J Antimicrob Chemother 63:269–273

4. Available on the centres of disease control website: CDC.gov using the web-link Disease Control website: CDC.gov. Accessed 31 Oct 2016. https://www.cdc.gov/pulsenet/pdf/ecoli-shigella-salmonella-pfge-protocol-508c.pdf

Chapter 13

Using RT qPCR for Quantifying *Mycobacteria marinum* from In Vitro and In Vivo Samples

Han Xaio and Stephen H. Gillespie

Abstract

Mycobacterium marinum, the causative agent of fish tuberculosis, is rarely a human pathogen causing a chronic skin infection. It is now wildely used as a model system in animal models, especially in zebra fish model, to study the pathology of tuberculosis and as a means of screening new anti-tuberculosis agent. To facilitate such research, quantifying the viable count of *M. marinum* bacteria is a crucial step. The main approach used currently is still by counting the number of colony forming units (cfu), a method that has been in place for almost 100 years. Though this method well established, understood and relatively easy to perform, it is time-consuming and labor-intensive. The result can be compromised by failure to grow effectively and the relationship between count and actual numbers is confused by clumping of the bacteria where a single colony is made from multiple organisms. More importantly, this method is not able to detect live but not cultivable bacteria, and there is increasing evidence that mycobacteria readily enter a "dormant" state which confounds the relationship between bacterial number in the host and the number detected in a cfu assay. DNA based PCR methods detect both living and dead organisms but here we describe a method, which utilizes species specific Taq-Man assay and RT-qPCR technology for quantifying the viable *M. marinum* bacterial load by detecting 16S ribosomal RNA (16S rRNA).

Key words Treatment monitoring, Antibiotic resistance, *Mycobacterium marinum*, Molecular diagnostics

1 Introduction

16 s rRNA, which accounts for 82–90% of the total RNA in myco-bacteria, is the core structural and functional component present in all bacteria. Its high abundance and critical functional makes it a suitable biomarker for mycobacterial quantification. Methods to detect *M. tuberculosis* have been described previously and applied successfully in clinical trials [1, 2]. This important observation has now been expanded as we have developed this assay further to make it more robust in laboratory practice and expanded the range of its use by designing species-specific Taq-man assay allowing the quantitative evaluation of *M. marinum* 16 s rRNA. Meanwhile, to take into account the potential loss of RNA during extraction

Stephen H. Gillespie (ed.), *Antibiotic Resistance Protocols*, Methods in Molecular Biology, vol. 1736,
https://doi.org/10.1007/978-1-4939-7638-6_13, © Springer Science+Business Media, LLC 2018

procedure, an internal control (IC) must be included. In this assay we use a fragment of potato RNA, with known concentration is spiked into the sample prior to RNA extraction to normalize such loss [1]. Taq-man assays for *M. marinum* 16 s rRNA and the IC are run simultaneously as a duplex qPCR run.

2   Materials

1. *Mycobacterium marinum* M strain (*see* **Note 1**).

2. Homogenization of the cell culture or tissues requires a homogenization kit—Microorganism lysing VK01-2 mL (Bertin Instrument), which contains 0.1 mm glass beads, is used for homogenization of the cells pellets or tissues.

3. RNA extraction: FastRNA PRO BLUE KIT (MP Biomedicals) or Purelink RNA mini kit (Invitrogen), are used for extraction of RNA (*see* **Note 2**).

4. DNA removal: DNA-free™ kit DNase (Invitrogen) Treatment and removal reagents are used for removing the DNA (*see* **Note 2**).

5. RT-qPCR: primers and probes should be purchased from a supplier using the sequences noted in Table 1 (*see* **Note 2**).

6. A QuantiTect-multiplex RT-PCR NR kit can be used to run the PCR (*see* **Note 2**).

7. A real time PCR machine, e.g., Rotor-Gene Q (Qiagen) (*see* **Note 3**).

8. Internal control: a segment of potato RNA is used as internal control, the generation of which is described in one of our papers published previously [1].

3   Method

3.1 Mycobacterial Quantification by Colony Forming Unit (CFU)

Carry out all procedures at room temperature unless otherwise specified.

Bacteria are quantified by a modified Miles and Misra method as described previously [3].

3.2 RNA Extraction and DNAse Treatment

3.2.1 Extraction of RNA from Liquid Culture of Mycobacterium marinum

1. Take 2 × 1 mL of the liquid culture, spin at $1,000,00 \times g$ for 10 min.

2. Remove the supernatant and resuspend the cell pellet in 950 μL of lysing buffer supplemented with 10% 2-mercaptoethanol for the Purelink RNA mini kit), which is provided by the RNA extraction kit.

Table 1
Taq-Man assay for *M. marinum* 16 s rRNA list of sequences for primers and probes

Name	Sequence	Channel	Target
M. marinum 16 s rRNA forward	5'-GAA CTC AAT AGT GTG TTT GGT GGT-3'		*Mycobacterium marinum* 16 s rRNA
M. marinum 16 s rRNA reverse	5'-ccc ATC CAA Aga cag GTG AA-3'		
M. marinum 16 s rRNA probe	FAM-TTG TCC GCC TCT TTT TCC CGT TT-BHQ1	Fam	
IC forward	5'-GTG TGA TAC TGT TGT TGA-3'		Internal control
IC reverse	5'-CCG Ata tag GGC TCT AAA-3'		
IC probe	Hex-TAC TCT CAG CCA CTA CCT CTC CAT-BHQ1	Hex	
Thermal cycles			
Step 1	50 °C 20 min	1 cycle	
Step2	94 °C 45 s	40 cycles	
	60 °C 45 s		

3. Spike in 50 ng of the internal control

4. Transfer the suspension to the homogenization tube and make sure it is tightly closed.

5. Place the homogenization tubes in the homogenizer and spin it using program 6.0 for 40 s if using Fastprep.

6. Transfer the homogenization tubes to a benchtop centrifuge and spin it at $12,000 \times g$ for 5 min.

7. Carefully transfer the supernatant to a clean tube without disturbing the glass beads.

8. A: If using Purelink RNA mini kit, follow the manufacturer's instruction by referring to the section of RNA Purification of the quick reference supplied with the kit.

9. B: If using FastRNA Pro, follow the manufacturer's instruction by referring to the quick reference protocol starting with step.

10. The extracted RNA could be subjected for DNase treatment immediately or stored at −20 °C if the DNase treatment is to be carried out within a month, or stored in −80 °C for future use.

3.2.2   RNA Extraction Using Zebrafish Embryos as an Example

1. Pool 10 or more embryos into a microcentrifuge tube and spin at $3000 \times g$ for 10 min.

2. Remove the supernatant without disturbing the embryo.

3. Add 950 μL lysing buffer supplemented with 10% 2-mercaptoethanol for the Purelink RNA mini kit (*see* **Note 4**).

4. Continue with **step 3** onward from Subheading 3.1.

3.2.3 DNAse Treatment

1. Make a master-mix of the Turbo DNase I 10× buffer and DNAse I enzyme for the number of samples plus 10% extra (*see* **Note 5**).

2. Mix by vortexing and then pipette 11 μL into each tube containing RNA extracted from Subheading 3.2.2.

3. Mix again by vortexing and then spin briefly (5–10 s at $13,000 \times g$).

4. Incubate at 37 °C for 30 min in the hot-block or incubator.

5. Add an additional 1 μL of DNase directly into each tube and mix well by vortexing.

6. Incubate at 37 °C for a further 30 min (*see* **Note 6**).

7. Thaw the DNase inactivation reagent 10 min prior the finish of DNase incubation and keep in the fridge. Resuspend by vortexing.

8. Add 10 μL of DNase inactivation reagent into each RNA extract.

9. Vortex three times during the 5-min incubation step at room temperature.

10. Centrifuge at $13,000 \times g$ for 2 min.

11. Transfer the supernatant to 1.5 mL RNase-free tube without touching any of the inactivation matrix.

3.3 RT-qPCR

1. Prepare 1 in 10 dilution of the RNA extracted from Subheading 3.2 in duplicate for RT-qPCR.

2. Prepare stock primer and probe with the final concentration as 10 μM.

3. Fluorescence signals are used as the read out of *M. marinum* 16 s rRNA and IC assay, are collected on Fam and Hex channel respectively (*see* **Note 7**).

4. Program the thermal cycler and include PCR reaction components as listed in Tables 1 and 2 respectively.

5. Make sure a no-RT reaction, for which reverse transcriptase is excluded for the RT-qPCR reaction components, is included for every sample to test if there is any DNA present in the sample.

Table 2
PCR reaction components

	RT+ reaction Volume per reaction (µL)	RT reaction Volume per reaction
QuantiTect mastermix	10	10 µL
M. marinum 16S F+ R primer mix	0.4	0.4 µL
M. marinum 16S–FAM probe	0.2	0.2 µL
IC F + R primer mix	0.4	0.4 µL
EC probe	0.2	0.2 µL
RT enzyme	0.2	–
Molecular grade water	4.6	4.8 µL
Sample	4	4
Total	20	20

3.4 Generation of the Correlations of CFU and Total RNA Detected by M. marinum 16 s rRNA Assay

1. Use liquid culture at late exponential phase.

2. Prepare seven decimal dilutions of the culture in triplicate.

3. Use one set of the dilutions to carry out a CFU counting and count the colony 5 days after the plating or once the colony is countable.

4. Use the duplicate dilutions prepared at **step 2** for RNA extraction, as described previously, and RT-qPCR.

5. A standard curve of M. marinum total RNA comprising 7 decimal dilutions with the highest concentration as 10 ng/µL–10^{-5} ng/µL.

6. Total RNA present in the sample prepared from **step 2** will be derived from the standard curve constructed from **step 5** (more information on data analysis can be found in Subheading 3.5).

7. Plot the CFU data against the corresponding amount of total RNA.

3.5 qPCR Data Interpretation and Bacterial Load Quantification

3.5.1 Principle

The principle of the MBL assay is absolute quantification based on a standard curve consisting of a set of RNA templates with known concentration. The standard curve is used to calculate the *M. marinum* concentration of an unknown sample.

IC standard curve is used to justify the efficiency of the extraction. If the amount of IC detected from unknown sample is no less than 10% of the spiked in IC, the extraction will be treated as a successful one, otherwise it will be treated as a failed extraction that should be repeated. Extraction efficiency could be achieved by divide the amount of IC from the sample by the spiked in IC, which can be used for normalization of the *M. marinum* MBL data. Standard curves must be constructed for each real-time PCR instrument (*see* **Note 8**).

3.5.2 Standard Curves Construction

1. *M. marinum* RNA extracted from culture with concentrations of 10^8 CFU/mL or higher and IC RNA at 50 ng/μL.

2. Dilute the extracted RNA decimally to create a series of standards. Add 10 μL of extracted RNA into 90 μL of RNase-free water, mix by vortexing for 5 s.

3. Set up the RT-PCR master mixes as outlined above in Table 2.

4. The standards are amplified in duplicates (along with the samples or on their own).

5. In RotorGene Q software, label the standards in sample sheet and assign them corresponding concentration and units, e.g., 10^8 for first 1 in 10 dilution (if the RNA is extracted from culture with 10^9 CFU/mL).

3.5.3 Standard Curve Data Analysis

The standard curve can be prepared in a separate run for the use with RotorGene Q and it can be further incorporated for data analysis of samples with unknown bacterial load.

1. Analyze the amplification curves in appropriate fluorescence channel, i.e., green channel for Mtb (FAM labeled probe), yellow channel for IC (VIC or HEX labeled probe).

2. Set the fluorescence threshold to 0.02 and examine the curves in exponential view and then in logarithmic mode.

3. Go to "Analysis" option and select the channel and sample sheet you are going to analyze.

4. Click on "Slope correct" in order to minimize the fluorescence fluctuations.

5. When standards and their respective concentrations are assigned in the sample sheet, the analysis software will automatically populate a standard curve.

6. Examine the parameters of the standard curve. The parameters are:

   (a) Slope (M), informs on assay efficiency.

   (b) Correlation coefficient (R^2), informs on assay linearity and the dynamic range (or limits of quantification).

   (c) Intercept, shift in C_T value on the y axis.

7. The PCR efficiency can be evaluated by the parameters of standard curve. The equation for an ideal standard curve and a 100% amplification efficiency ($E = 1$) is:
   $$C_T = slope \times Log(concentration) - intercept.$$

Table 3
Validation of assay

Target (Marinum)	IC	Result
+	+	+
+	−	+*
−	+	−
−	−	Invalid

+ = Positive shown by Cycle threshold (Ct) from the RT-qPCR
− = Negative shown by no Ct from the RT-PCR
* = The Mtb presence result is positive, but the result cannot be used for quantitative analysis or data normalization

or
$$C_T = -3.32 \times \text{Log(concentration)} - \text{intercept}.$$
Aim for the efficiency of 90–100%, i.e., $E = 0.9$–1.0. The efficiency can be calculated from the slope of the standard curve using the equation:
$$E = 10^{-1/-3.32} - 1.$$

8. Very high or too low RNA concentrations in the RT-PCR reaction can cause fluctuations in reverse transcription and PCR efficiency. These result in outlying C_T values. Outlier C_T values can be also caused by errors in pipetting, dilutions' preparation, and insufficient homogeneity of a PCR master-mix, evaporation during reaction and improperly placed rotor.

9. Consider careful removal of the outliers.

10. Interpretation of the data is illustrated in Table 3.

4   Notes

1. The *M. marinum* can be grown in Middlebrook 7H9 broth supplemented with OADC and incubated at 30 °C.

2. The method for RNA extraction DNA digestion and PCR master mix presented in this chapter is optimized using the kits noted but alternatives can be used.

3. The assay presented in this chapter is optimized for the Rotor-Gene (Qiagen) but other machines with similar characteristics can be used and we have adapted similar assays to a wide range of machines.

4. After this procedure the material can be stored at −80 °C.

5. After defrosting of the Multiplex QuantiTect master mix, aliquot them in 500 μL and store them at −20 °C. Avoid multiple freeze and thaw of the master mix which can reduce its efficiency. If the mix is not finished at a single use, it can be stored at 4 °C for up to a week for further use.

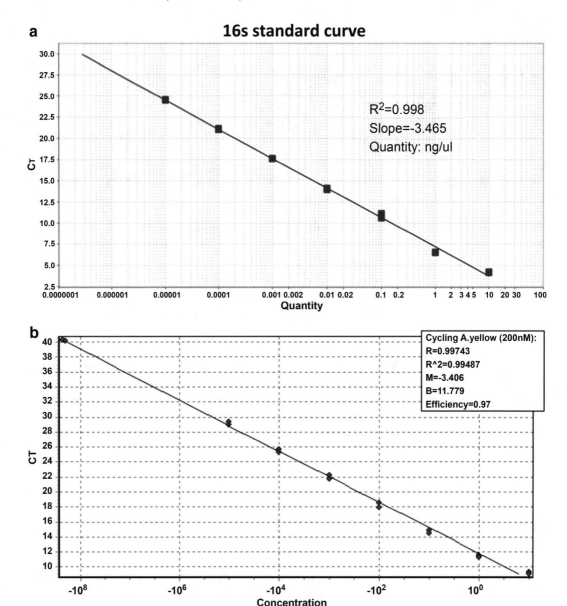

Fig. 1 (**a**) Standard curve of M. marinum 16s rRNA, (**b**) standard curve of IC

6. Incubation time can be up to 1 h.

7. Optimization has been carried out to ensure that there is no cross-reaction between these two reactions.

8. Extraction efficiency can be achieved by dividing the IC in the post extraction sample by the spiked in amount. Actual 16 s rRNA present in the preextraction sample can therefore be retrieved based on the extraction efficiency of each individual sample. Efficiency of *M. marinum* 16 s rRNA (Fig. 1a) and IC (Fig. 1b) Taq-man assay was tested by running a standard

curve composing pure RNA with six decimal dilutions, with the highest concentration of 10 ng/μL while the lowest concentration of 10^{-5} ng/μL. Based on three replicate experiments, efficiencies of these two assays are above 95%. To reflect the CFU of *M. marinum* from each sample, correlation between CFU and the amount of total RNA detected by 16 s rRNA assay was established. However, such correlation is used only as an indicator of the approximate amount of CFU present in the sample but not to conclude the actual amount of CFU. Such recommendation is based on the observation that relationship between CFU and 16 s rRNA varies among different growth phase and when under different stress conditions.

Acknowledgments

The research leading to this chapter was supported by the Innovative Medicines Initiative Joint Undertaking under grant agreement no 115337, resources of which are composed of financial contribution from the European Union's Seventh Framework Programme (FP7/2007-2013) and EFPIA companies' in-kind contribution.

References

1. Honeyborne I, McHugh TD, Phillips PPJ et al (2011) Molecular bacterial load assay, a culture-free biomarker for rapid and accurate quantification of sputum mycobacterium tuberculosis bacillary load during treatment. J Clin Microbiol 49:3905–3911

2. Honeyborne I, Mtafya B, Phillips PPJ et al (2014) The molecular bacterial load assay replaces solid culture for measuring early bactericidal response to antituberculosis treatment. J Clin Microbiol 52:3064–3067

3. Billington OJ, McHugh TD, Gillespie SH (1999) Physiological cost of rifampin resistance induced in vitro in *Mycobacterium tuberculosis*. Antimicrob Agents Chemother 43:1866

Chapter 14

Use of Larval Zebrafish Model to Study Within-Host Infection Dynamics

Tomasz K. Prajsnar, Gareth McVicker, Alexander Williams,
Stephen A. Renshaw, and Simon J. Foster

Abstract

Investigating bacterial dynamics within the infected host has proved very useful for understanding mechanisms of pathogenesis. Here we present the protocols we use to study bacterial dynamics within infected embryonic zebrafish. This chapter encompasses basic techniques used to study bacterial infection within larval zebrafish, including embryonic zebrafish maintenance, injections of morpholino oligonucleotides, intravenous injections of bacterial suspensions, and fluorescence imaging of infected zebrafish. Specific methods for studying bacterial within-host population dynamics are also described.

Key words Zebrafish, Infection, Bacterial population dynamics, Fluorescence microscopy

1 Introduction

Animal models of human infection have proven an effective way to elucidate the mechanisms of bacterial pathogenesis. Zebrafish are commonly used in bacterial pathogenesis studies [1] mainly due to their small size, ease of breeding, and similarities of their immune system components to those of humans [2]. Macrophages and neutrophils are already present in developing zebrafish at 30 h postfertilization (hpf). In addition, the optical transparency of embryonic and larval zebrafish allows visualization of immune cell types interacting with invading pathogens in real time. Both host and pathogen cells can be fluorescently labeled either genetically or with fluorescent dyes [3, 4] to visualize pathogen subcellular localization as well as the pH in their intraphagocyte milieu [4].

Additionally, morpholino-modified antisense oligonucleotides (morpholinos or MOs) have been extensively used in zebrafish to knock down genes of interest by either blocking the translation or interfering with the RNA splicing process. MO-mediated knockdown can be easily used as a measure to study the influence of host factors on the bacterial infection process. We would stress the

Stephen H. Gillespie (ed.), *Antibiotic Resistance Protocols*, Methods in Molecular Biology, vol. 1736,
https://doi.org/10.1007/978-1-4939-7638-6_14, © Springer Science+Business Media, LLC 2018

importance of confirming morpholino findings with mutant experiments [5].

Recently, several methods have been established to understand the pathogen dynamics within an infected host. Multiple genetically labeled wild-type isogenic tagged strains (WITS) have been successfully applied to modeling the dynamics of *Salmonella enterica* infection. This approach has shown a great potential to increase the understanding of the mechanisms that underpin infection processes [6]. Eight WITS simultaneously injected into a single host permitted the enhanced resolution of bacterial subpopulation tracing and dynamics in vivo. Alternatively, isogenic strains labeled with different antibiotic resistance markers have been used for population dynamic studies [7, 8]. The use of genetically tagged WITS requires laborious molecular biology techniques such as qPCR, whereas our antibiotic resistant isogenic strains can be quantified by simple plating on selective media. Additionally, antibiotic resistance enables investigation into the effects of antibiotic therapy on bacterial population dynamics [8].

In this chapter, we describe in detail the techniques used to investigate pathogenesis, host–pathogen interaction and bacterial within-host population dynamics using the zebrafish model of infection.

2   Materials

1. E3 larval zebrafish medium (×10): 50 mM NaCl, 1.7 mM KCl, 3.3 mM $CaCl_2$, 3.3 mM $MgSO_4$. Prepare the 10× stock and subsequently dilute to 1× solution in distilled water. In order to prevent fungal growth, supplement the E3 medium with Methylene Blue to a final concentration of 0.00005% (w/v, approximately four drops of 0.05% Methylene Blue per litre). Autoclave and cool to approximately 28 °C before use.

2. Tricaine (zebrafish anesthetic). Prepare a stock solution of 0.4% (w/v) 3-amino benzoic acid ester (tricaine or MS322) in 20 mM Tris–HCl, adjust the pH to 7.0 and store at −20 °C (*see* **Note 1**).

3. Methylcellulose solution. Prepare the 3% (w/v) methylcellulose solution in E3 (with Methylene Blue). Aliquot the clarified solution into 20 mL syringes and freeze for long-term storage. For use and short-term storage, keep in the zebrafish incubator (*see* **Note 2**).

4. Low-melting point (LMP) agarose for mounting embryos prior to microscopic imaging. Prepare 0.5% (w/v) LMP agarose in E3 medium (without Methylene Blue), heat up the

suspension to solubilize the agarose and cool in a 37 °C water bath before use.

5. Dissecting stereomicroscope with transmitted illumination, providing at least 50× magnification with adjustable zoom objective.

6. A pair of Dumont #5 watchmaker forceps with "standard" tip (*see* **Note 3**).

7. 3 mL graduated Pasteur pipettes for embryo transfer (*see* **Note 4**).

8. Mineral oil for scaling injection drop size.

9. Calibration slide with 0.05 or 0.1 mm intervals.

10. Concentrated bleach.

11. Standard microscope slides.

12. 1 mm diameter capillary tubes without filaments, 10 cm long.

13. Micropipette puller (*see* **Note 5**).

14. Pneumatic microinjector with micromanipulator.

15. Fluorescence microscope with 4× air, 10× air, and 60× oil objectives.

16. Glass bottom dish for inverted microscopy (*see* **Note 6**).

17. 100 mm petri dishes and 96-well plates with lids.

18. Suitable centrifuge tubes for bacterial suspension handling and centrifugation (1.5 and 50 mL).

3   Methods

Figure 1 shows a timeline of embryonic zebrafish infection experimental procedure involving adult fish mating, collecting eggs, possible MO injections, culturing of bacterial strains, injections into zebrafish blood circulation, and following the experiment until 5 days post-fertilization (dpf).

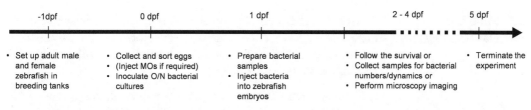

Fig. 1 Timeline of embryonic zebrafish infection experimental procedure. The timeline represents the typical zebrafish infection experiment, starting from setting up adult fish for eggs until terminating the experiment at 5 dpf (legal protection age in the UK when grown at 28 °C)

3.1 Maintenance of Embryonic Zebrafish

1. Collect freshly laid embryos from the fish tank into a petri dish (*see* **Note 7**).

2. Decant aquarium water from the dish and refill plate with fresh E3 solution.

3. Transfer embryos into fresh petri dishes (approximately 60 embryos per dish), use a stereomicroscope to ensure embryos are at the similar developmental stage.

4. Top up with E3 and place petri dishes into incubator, kept at a constant 28 °C.

3.2 Morpholino Injections

Microinjections of morpholinos are performed using a pneumatic micropump, a micromanipulator and a dissecting microscope (similarly to *S. aureus* intravenous injections).

1. Collect zebrafish embryos from aquarium facility promptly after eggs being laid to obtain eggs at no later than 1–4 cell stage.

2. Load a glass injection needle with approximately 3 µL of morpholino solution using 20 µL microloader pipette tip.

3. Mount the needle on the injector nozzle installed on the micromanipulator.

4. Break the needle to ideally form oblique angled tip (*see* **Note 8**).

5. Calibrate the droplet size using a calibration slide (*see* **Note 9**).

6. Using an edge of microscopic slide placed inside a petri dish lid, line up approximately 50 embryos against the side of the slide and remove excess liquid.

7. Inject 0.5–1 nL of morpholino solution into the center of embryo yolk (*see* **Note 9**).

8. After injections, remove the lined-up embryos from the microscopic slide edge by gently applying E3 using a Pasteur pipette.

9. Transfer injected eggs into a fresh petri dish.

10. Repeat **steps 5–8**, depending on number of knockdown embryos required (*see* **Note 10**).

3.3 Preparation of Bacterial Inoculum

For within-host bacterial population studies use wild-type isogenic strains labeled with either different antibiotic markers (construction described in McVicker et al. submitted protocol) or with different fluorescence markers [6].

1. Grow an overnight starter culture by inoculating a few bacterial colonies into in 5–10 mL of BHI and placing into 37 °C with 250 rpm orbital shaking (*see* **Note 11**).

2. Inoculate 50 mL of fresh BHI with the 0.5 mL of starter culture (1:100) and grow for 2 h until the optical density at 600 nm (OD_{600}) reaches approximately 1 (*see* **Note 12**).

3. Transfer 40 mL of culture into 50 mL centrifuge tube, and spin down at $5000 \times g$ for 10 min at 4 °C.

4. Based on OD_{600} readings, resuspend samples in appropriate amount of PBS (*see* **Note 12**) to achieve bacterial concentrations of 1.5×10^9 CFU/mL.

5. Mix together an appropriate amount of each strain on ice to create a suspension of bacteria at the ratio required by your experiment (usually 1:1:1). Vortex well, both before and after mixing.

3.4 Preinfection Preparation of Zebrafish Embryos

1. When zebrafish embryos reach approximately 28 hpf, using a pair of forceps, dechorionate them manually under the stereomicroscope (*see* **Note 13**). Remove empty chorions from the petri dish afterward.

2. 5–10 min prior to bacterial injections at 30 hpf, anesthetize zebrafish embryos by addition of tricaine to a final concentration of 0.02% (w/v). Due to photosensitivity of tricaine, cover the dish immediately with a dark object to prevent exposure to light.

3. Apply an approx. 1 cm wide line of methylcellulose onto a microscope slide from one end to the other.

4. Collect the anesthetized zebrafish embryos into a 3 mL Pasteur pipette and place approximately 30 embryos along the line of methylcellulose (*see* **Note 14**). Remove any excess liquid water with paper tissue.

3.5 Bacterial Injections into Zebrafish Embryos

1. Load a glass injection needle with approximately 5 µL of previously prepared bacterial suspension using 20 µL microloader pipette tip.

2. Mount the needle on the injector nozzle installed on the micromanipulator.

3. Break the needle tip to ideally to form oblique angled tip (*see* **Note 8**).

4. Calibrate the droplet size using a calibration slide.

5. Place the slide with mounted fish on methylcellulose onto the injecting stage.

6. Inject a number of pulses into 1 mL of PBS for dose control (*see* **Note 15**) prior to zebrafish injections.

7. Inject bacteria into zebrafish yolk sac circulation valley (*see* Fig. 2a and **Note 16**).

8. After injecting a set of embryos repeat **step 5** for dose control at the end of the set.

9. Remove the injected set of embryos from the slide, remove embryos from methylcellulose (*see* **Note 17**) and transfer them into E3 solution in a petri dish.

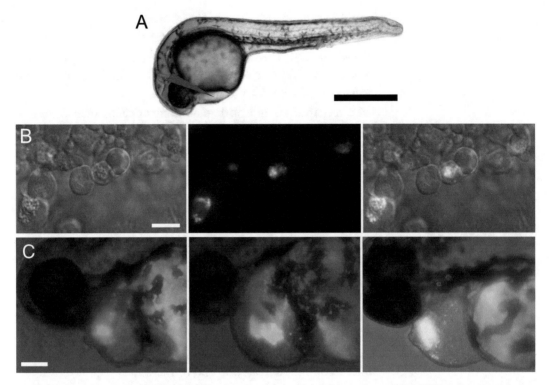

Fig. 2 Microscopy using zebrafish. (**a**). Microscopic image of zebrafish embryo at 30 hpf. The site and direction of bacterial injection into the circulation is indicated by the red arrow. Scale bar indicates 500 μm. (**b**). In vivo image of the yolk circulation valley of 32 hpf larvae, 2 h after injection with a mixture consisting of 750 CFU of CFP-labeled (Cyan Fluorescent Protein) and 750 CFU of YFP-lablled (Yellow Fluorescent Protein) *S. aureus*. Left panel shows DIC, middle panel—fluorescence, and right panel—merge image. Two differently labeled bacterial strains were equally distributed within phagocytes at early stages of infection. Scale bar indicates 10 μm. (**c**). Examples of In vivo images of terminally infected wild-type larvae (24–44 hpi) after injection with a mixture consisting of 750 CFU of CFP-labeled and 750 CFU of YFP-labeled *S. aureus*. At later stages of infection, each abscess-like structure seen was formed almost exclusively by bacteria with a single fluorescent label (CFP-labeled bacteria for the left panel, and YFP-labeled bacteria for the right panel). In the middle panel, lesions are formed by two bacterial strains, but they are spatially separated. Scale bar indicates 100 μm

10. Return plate to incubator and leave to "rest" for 90 min.

11. Pour off E3 and refill with fresh E3 (*see* **Note 18**).

12. Transfer embryos into a 96-well plate (each embryo to a separate well) topping up well to about 3/4 full with E3 (for survival and microscopy experiments).

13. In order to verify the inoculum dose, spot 10 μL of injected samples of 1 mL of PBS (*see* Subheading 3.5, **steps 6** and **8**) into a BHI agar plate (in triplicate).

14. Observe infected zebrafish embryos at intervals for the end-point of diminished blood circulation, visually identifiable bacterial lesions and collect for either determination of in vivo bacterial numbers or fluorescence microscopy.

3.6 Determination of In Vivo Total and Relative Bacterial Loads

For bacterial population dynamics experiments, zebrafish embryos should be infected as normal, but care should be taken to place infected fish with exact amount of E3 (e.g., 100 μL, *see* **Note 19**).

1. Transfer infected embryos individually into homogenizer tubes together with 100 μL of E3 from a well at different time intervals (if time-course experiment is required), or while the embryo is heavily infected/dead (for within-host population dynamics).

2. Freeze collected samples at −20 °C.

3. After collecting all samples, defrost them for approximately 30 min prior to use.

4. Homogenize defrosted embryos using a homogenizer, e.g., PreCellys 24-Dual. Typical homogenization time is 30 s.

5. Serially dilute up to 3 orders of magnitude ("neat" plus 3 decimal dilutions) using 96-well plates and 8-channel pipette (*see* **Note 20**).

6. Plate out 5 μL starting with a row of most diluted samples onto BHI agar supplemented with appropriate antibiotics.

7. Incubate the agar plates overnight at 37 °C.

8. Count colonies of convenient dilution and determine number of CFU per embryo depending on dilution factor used for counting colonies.

3.7 Imaging of S. aureus-Infected Zebrafish In Vivo Using Fluorescence Microscopy

1. Prepare 1% (w/v) low-melting point agarose solution in E3 medium (without Methylene blue) (*see* **Note 21**).

2. Anesthetize embryos using tricaine as instructed previously (*see* Subheading 3.4).

3. Place embryos into a glass bottom dish and immerse embryos in low-melting point agarose solution (*see* **Note 22**).

4. Wait for approximately 5 min (until the agarose solidifies) and cover embedded embryos with E3 solution supplemented with tricaine.

5. Use 2× objectives when imaging the entire embryo body (*see* Fig. 2a), 10× for large section of embryos (*see* Fig. 2c), and 60× for imaging bacterial and host cellular interactions (*see* Fig. 2b).

3.8 Data Presentation and Statistical Analysis

GraphPad Prism graphing and statistics software is commonly used to present and analyze data.

1. For presenting survival curves and comparing virulence of different bacterial strains, use a Kaplan–Meier survival plot, and perform a pair-wise statistical analysis using Log-rank test. A Bonferroni correction is recommended when performing mul-

tiple comparisons (e.g., a range of mutants tested versus one control parental strain).

2. For bacterial numbers recovered from infected embryos Mann–Whitney test is recommended as the numbers do not follow the Gaussian distribution (two categories of low bacterial numbers within surviving embryos and high bacterial numbers which succumb to infection).

4 Notes

1. Working stock of tricaine can be kept at room temperature, but since tricaine is light-sensitive, it needs to be kept in the dark (the container should be wrapped in tin foil). In addition, a petri dish with anesthetized zebrafish embryos should be also covered to protect it from direct light.

2. To facilitate methylcellulose solubilization, use several rounds of partial freezing and thawing to dissolve the white clumps of methylcellulose. Freeze for approximately 30 min (until layer of ice is formed on the surface) and mix thoroughly while thawing. Repeat this action for around five times.

3. We do not recommend the use of forceps with very sharp tips (e.g., "biology" type in Fine Science Tools catalogue) as they tend to damage zebrafish embryos while dechorionating (*see* Subheading 3.4).

4. Use pipettes with an opening of approx. 2 mm, as smaller size could be damaging to zebrafish embryos while transferring.

5. In order to achieve optimal needle thickness and shape for successful injections, establishing the correct parameters on the needle puller is essential. Use thin and long needles for intravenous injections. Our settings on a Sutter Instrument Model P-97 puller are: Heat 430, Pull 225, Velocity 150, Time 225.

6. The glass bottom should be very thin (ideally glass thickness No. 0) to enable high magnification imaging using 60× oil objective.

7. For morpholino injections, zebrafish embryos have to be collected very early (within half an hour of laying), to make sure injections are performed before embryos reach the 4-cell stage. If MO injections are not required, embryos can be collected up to a few hours later.

8. Using the dissecting microscope set on the highest magnification (e.g., 50×), gently scrape the tip of the glass needle with forceps. Ensure that only small amount of the needle tip is broken off.

9. Volumes of 0.5–1 nL are typically used and morpholino concentrations vary between 200 μM and 1 mM to effectively knock down gene activity without causing off-target effects. 0.5% (v/v) solution of Phenol Red can be added to the morpholino solution (at 1:10 ratio) to "colorize" the injecting material and therefore facilitate the injections. Same dye can be added to bacterial suspensions.

10. Mortality rates of MO-injected embryos may reach 50%, therefore inject double the number required.

11. There is no need to use antibiotics if genomic markers are used. Bacteria with antibiotic markers located on a plasmid still require supplementing appropriate antibiotics in liquid media.

12. OD of 1 at 600 nm corresponds to bacterial concentration of approximately 2×10^8 CFU/mL.

13. Carefully pinch one side of the fish chorion with one hand and 'peel' it open with the other. If fish is damaged in any way, remove immediately into bleach solution (should be prepared in a bucket near your setup).

14. Very carefully decant the embryos out onto a line of methylcellulose. Do not break the surface tension between pipette and methylcellulose. Slowly drag the pipette back across the line of methylcellulose. This should separate the fish out from the initial drop you made. Use this method all the way along the line and you should have a well dispersed line of fish.

15. Prior to testing the dose control, ensure needle is unblocked and is producing visible suspension by ejecting into methylcellulose. Number of pulses injected into the PBS solution should depend on expected bacterial counts.

16. The needle should approach the embryo from the dorsolateral side (*see* Fig. 2a). Between injections into embryos, keep ensuring that needle is unblocked and is producing visible suspension by ejecting into methylcellulose.

17. To remove zebrafish embryos from methylcellulose use a Pasteur pipette to place a few drops of E3 onto the line of fish. Carefully aspirate the fish into the pipette. Try not to bring any air bubbles with the injected zebrafish. Carefully place the fish back into a petri dish with fresh E3 solution, pumping up and down slowly to aid their removal from the pipette.

18. Check whether embryos are not still stuck in viscous methylcellulose material. If they still are, use a pipette to gently stream some E3 over them to remove them.

19. In experiments regarding determination of bacterial numbers within infected embryos, it is important to collect an embryo with a known amount of E3 medium (for subsequent serial dilution factors and colony counts). It is convenient, therefore,

to place freshly infected embryos in a set volume of E3 (e.g., 100 μL) within 96-well plates wells. It is also advisable to place the plates into moist environment (e.g., a box with wet tissue paper) to prevent evaporation, and therefore keeping the volume constant throughout the experiment.

20. The total number of bacteria within an embryo at the end-point of successful infection is approximately 10^6 CFU per embryo.

21. Dissolve low-melting point agarose by heating up the suspension and cool down to 37 °C before use.

22. It is important to place the embryos as flat against the glass as possible to minimize optical distance between the sample and the microscope lens. Lateral orientation of embryos is preferred while imaging.

References

1. Torraca V, Masud S, Spaink HP, Meijer AH (2014) Macrophage-pathogen interactions in infectious diseases: new therapeutic insights from the zebrafish host model. Dis Model Mech 7:785–797

2. Henry KM, Loynes CA, Whyte MK, Renshaw SA (2013) Zebrafish as a model for the study of neutrophil biology. J Leukoc Biol 94:633–642

3. Prajsnar TK, Cunliffe VT, Foster SJ, Renshaw SA (2008) A novel vertebrate model of *Staphylococcus aureus* infection reveals phagocyte-dependent resistance of zebrafish to non-host specialized pathogens. Cell Microbiol 10:2312–2325

4. Prajsnar TK, Renshaw SA, Ogryzko NV, Foster SJ, Serror P, Mesnage S (2013) Zebrafish as a novel vertebrate model to dissect enterococcal pathogenesis. Infect Immun 81:4271–4279

5. Schulte-Merker S, Stainier DYR (2014) Out with the old, in with the new: reassessing morpholino knockdowns in light of genome editing technology. Development 141:3103–3104

6. Grant AJ, Restif O, McKinley TJ, Sheppard M, Maskell DJ, Mastroeni P (2008) Modelling within-host spatiotemporal dynamics of invasive bacterial disease. PLoS Biol 6:e74

7. Prajsnar TK, Hamilton R, Garcia-Lara J, McVicker G, Williams A, Boots M, Foster SJ, Renshaw SA (2012) A privileged intraphagocyte niche is responsible for disseminated infection of *Staphylococcus aureus* in a zebrafish model. Cell Microbiol 14:1600–1619

8. McVicker G, Prajsnar TK, Williams A, Wagner NL, Boots M, Renshaw SA, Foster SJ (2014) Clonal expansion during *Staphylococcus aureus* infection dynamics reveals the effect of antibiotic intervention. PLoS Pathog 10:e1003959

Chapter 15

A Method to Evaluate Persistent *Mycobacterium tuberculosis* In Vitro and in the Cornell Mouse Model of Tuberculosis

Yanmin Hu and Anthony Coates

Abstract

Persistent *Mycobacterium tuberculosis* will not grow on solid or liquid media. They will, however, grow in the presence of resuscitation promoting factors (RPF). Here we describe the production of RPF rich culture supernatants, and their use for the stimulation of growth of persisters in vitro as well as in the Cornell model of tuberculosis.

Key words Persisters, *Mycobacterium tuberculosis*, Resuscitation promoting factors, Culture supernatants, Cornell model

1 Introduction

Culture of bacteria in vitro has been the pillar of diagnosis and research in microbiology since the days of Pasteur and Koch. These techniques are more sensitive than microscopy alone and provide large numbers of bacteria for subsequent antibiotic susceptibility testing and characterization.

Traditionally, viable bacterial enumeration is based on colony forming unit (CFU) counts on agar plates or broth counts in liquid medium. Testing and development of novel anti-TB drug regimens mainly focus on therapeutic endpoints based on sputum negative CFU counts or no mycobacterial growth in broth. However, observation of the treatment of tuberculosis in humans [1] and in experimental animals [2, 3] provides strong evidence that nonmultiplying persisters of *M. tuberculosis* are present at the beginning and enriched during treatment. Even after 6 months of chemotherapy, persisters are still present and cause subsequent disease relapse. These persistent bacteria [4] remain undetected by conventional culture methods. They do not retain acid-fast stains or multiply on agar or in broth culture medium, which makes their

Stephen H. Gillespie (ed.), *Antibiotic Resistance Protocols*, Methods in Molecular Biology, vol. 1736,
https://doi.org/10.1007/978-1-4939-7638-6_15, © Springer Science+Business Media, LLC 2018

detection extremely challenging for clinical studies, and explains the paucity of effective therapeutic agents. These "invisible" persistent bacteria, therefore, represent an underexplored therapeutic target, which if successfully detected and eliminated, would not only shorten anti-TB chemotherapy duration but also reduce disease relapse.

Essentially, to eliminate these persisters, one must first detect them. One of the most intuitive and promising methods is to "wake up" the persistent bacteria from their dormant state, and induce them to recommence multiplication. Previous studies have showed that culture supernatant from *M. tuberculosis* young culture contains resuscitation promoting factors (RPF), 5 secreted proteins RPF-A, RPF-B, RPF-C, RPF-D, and RPF-E. These RPF proteins are able to stimulate persistent bacteria to initiate multiplication [5, 6]. This generates bacterial counts from conventionally culture-negative samples, which enables quantitative persister detection [5, 7]. In our recent study [7], we demonstrated that persistent *M. tuberculosis* which depended on culture supernatant to replicate were present in stationary phase cultures in vitro and in *M. tuberculosis* infected mouse organs. The long duration of tuberculosis treatment is due to the presence of RPF-dependent persisters [8]. We also demonstrated for the first time that high-dose rifampicin drug regimen was able to kill RPF-dependent persistent bacteria, enabling a shortened treatment duration in mice without subsequent disease relapse [7].

M. tuberculosis rpf-like genes are expressed in exponential growth phase in vitro [9, 10], in infected animals [10] and in humans [11]. Using self-generated *M. tuberculosis* RPF to stimulate persistent bacteria is accurate and efficient with reproducible experimental results [7, 8]. Culture supernatant containing RPF are essentially collected from exponential growth phase cultures which can be achieved under aerobic and microaerophilic conditions [5, 7]. The mechanisms by which RPF proteins resuscitate dormant cells remain unknown although peptidoglycan hydrolysis by RPF has been proposed [12].

In this chapter, we describe methods for the production of *M. tuberculosis* culture filtrates containing RPF and resuscitation of RPF-dependent persisters in an in vitro hypoxic model and in the Cornell model of tuberculosis.

2 Materials

2.1 Growth and Preparation of Mycobacteria

1. Middlebrook 7H9 Broth: add 4.7 g of powder 7H9 medium (Becton Dickinson, UK), 2 mL of glycerol and Tween 80 to the final concentration of 0.05% (v/v) to 900 mL of distilled water (*see* **Note 1**) followed by autoclaving at 121 °C for 15 min. After sterilization, cool the medium to room

temperature (*see* **Note 2**) and add 100 mL of Albumin Dextrose Complex (ADC) (*see* Subheading 2.1, **item 7**).

2. Middlebrook 7H11 Agar: Add 21 g of powder 7H11 medium (Becton Dickinson, UK) and 5 mL of glycerol to 900 mL of distilled water followed by autoclaving at 121 °C for 15 min. After sterilization, cool the agar to 55 °C (*see* **Note 2**) and add 100 mL of Oleic Albumin Dextrose Complex (OADC, Becton Dickinson, UK) (*see* Subheading 2.1, **item 8**) immediately before pouring to 90 mm sterile petri dishes.

3. Kirchner liquid medium: add 17.4 g of powder Kirchner liquid medium (MAST) and 20 mL of glycerol to 1 L of distilled water, autoclave at 121 °C for 15 min. After sterilization, cool the medium to room temperature and add 100 mL of sterile inactivated horse serum (Oxoid, UK) before use.

4. Löwenstein-Jensen slopes: In 600 mL purified water, add monopotassium phosphate 2.4 g, potato flour 30 g, magnesium sulfate 0.24 g, asparagine 3.6 g, sodium citrate 0.6 g, and Malachite Green 0.4 g and glycerol 12 mL. After sterilization by autoclaving at 121 °C for 15 min, add fresh 1000 mL whole egg. Dispense the complete medium into sterile screw capped tubes and arrange in a slant position in a suitable rack. These slopes are obtained from Becton Dickinson and stored at 2–8 °C in the dark (*see* **Note 3**).

5. Blood agar: Add 39 g of powder Columbia blood agar base (Oxoid, UK) to 1 L of distilled water. Sterilize by autoclaving at 121 °C for 15 min. Cool to 50 °C and add 5% sterile defibrinated blood (Oxoid) immediately before pouring to 90 mm sterile petri dishes.

6. Sabouraud dextrose agar: Add 65 g of Sabouraud dextrose agar (Oxoid) to 1 L of distilled water. Sterilize by autoclaving at 121 °C for 15 min. Mix well and pour into 90 mm sterile petri dishes.

7. Albumin Dextrose Complex (ADC): add 10 g of albumin fraction V, 4 g of D-glucose, and 1.7 g NaCI to 200 mL sterilized distilled water. Stir to completely solubilize the albumin. Sterilize using 0.2 μm fliter and store at 4 °C.

8. Oleic Albumin Dextrose Complex (OADC): obtain from Becton Dickinson, UK. OADC contains 5% of albumin fraction V, 2% of glucose, 0.85% of NaCI, 0.05% of oleic acid, and 0.004% catalase.

9. Phosphate Buffered Saline: NaCl 0.138 M, KCl 0.0027 M, pH 7.4.

10. Selectatab (Mast Diagnostica GmbH): each tablet contains polymyof polymyxin B 200,000 unit/L, carbenicillin 100 mg/L, trimethoprim 10 mg/L, and amphotericin B 10 mg/L.

2.2 Cornell mouse Model

1. Mice: BALB/c mice, female and aged 6–8 weeks (Harlan UK Ltd.).

2. Rifampicin, isoniazid and pyrazinamide are obtained from Sigma-Aldrich.

3. Hydrocortisone acetate is obtained from Sigma-Aldrich.

3   Method

As transmission of *M. tuberculosis* is via aerosols, workers must be protected from laboratory acquired infections caused by *M. tuberculosis*. Working with *M. tuberculosis* must be in Biosafety cabinet (BC) within the Biosafety level 3 laboratories.

3.1 Preparation of Culture Filtrates Containing RPF

Culture supernatant containing resuscitation promoting factors (RPF) or 7H9 medium is used as described previously [5, 7, 8].

1. *M. tuberculosis* H37Rv is grown in serials of 10 mL 7H9 broth media in 30 mL screw-capped universal tubes without disturbance (*see* Subheading 3.2) for 15–20 days until an optical density of 1–1.5 (optical density reader, Biochrom WPA CO8000) is reached (*see* **Note 4**).

2. The bacteria are removed by centrifugation at 3000 × *g* for 15 min.

3. The culture supernatants are collected and sterilized by filtration with 0.2 µm filter (Sartorius) twice.

4. The sterilized culture filtrates are made selective by addition of Selectatab (Mast Diagnostica GmbH) (*see* Subheading 2.1, **item 11**).

5. The selective culture filtrates are used immediately for broth counting of the most probable number (MPN) of the bacilli (*see* Subheading 3.6).

3.2 In Vitro Hypoxia Model of M. tuberculosis Growth

1. Serials of 10 mL 7H9 broth media in 30 mL screw-capped universal tubes are inoculated with 1 mL of 10-day *M. tuberculosis* culture (*see* **Note 5**).

2. The cultures are incubated in an upright position without disturbing for up to 100 days.

3. The numbers of viable *M. tuberculosis* in the cultures are determined by surface plate counts on 7H11 agar.

4. Prior to inoculation, the agar plates are incubated upside down at 37 °C for 24 h in order to check sterility and to ensure the surface is sufficiently dry.

5. Cultures are vortexing in 30 mL screw-capped bottles with 1 mm glass beads (VWR UK) for 2–5 min (*see* **Note 6**).

6. Place the culture in an ultrasonic water bath (Branson Ultrasonic B. V.) for 5 min in order to obtain uniformly dispersed single cell suspension (*see* **Note 7**).

7. Serials of tenfold dilutions of the cultures are made in 7H9 broth with 0.05% (v/v) Tween 80 but without ADC.

8. 100 μL of samples are added to one-third segments of the agar plates in duplicate.

9. The inocula are allowed to dry into the agar and the plates are incubated in double polythene bags for 3 weeks at 37 °C.

10. Viability is expressed as colony forming units (CFU) per milliliter (*see* **Note 8**).

3.3 Antibiotic Exposure In Vitro

1. Antibiotics at different concentrations are added into log-phase and stationary-phase cultures in the hypoxia model (*see* Subheading 3.2).

2. The cultures are incubated at 37 °C without disturbance.

3. At different time point, the cultures are washed with phosphate buffered saline (PBS) for three times to remove the antibiotics.

4. Viability is determined using CFU counting (*see* Subheading 3.2) or broth counting (*see* Subheading 3.6).

3.4 Cornell mouse Model

1. BALB/c mice are infected intravenously via the tail vein with 1.2×10^5 CFU of mouse-passaged *M. tuberculosis* strain H37Rv per mouse [13, 14] (*see* **Note 9**).

2. Mice are randomly allocated into experimental groups and control group.

3. Control group consists of infected and untreated mice.

4. Four of these are sacrificed at 2 h after infection to monitor initial bacterial loading in lungs and spleens of mice.

5. Four are killed at the beginning of treatment, 3 weeks after infection (*see* **Note 10**).

6. The treatment groups are administrated with a combination of rifampicin (R), isoniazid (H), and pyrazinamide (Z) for 16 weeks (*see* **Note 11**).

7. Treatment is given by daily gavage (0.2 mL) for 5 days per week at the dosages of R 10 mg/kg, H 25 mg/kg, and Z 150 mg/kg.

8. The drug suspension is prepared freshly for the daily dosage.

9. Immediately after termination of 16 weeks of chemotherapy, the remaining mice are administered 0.5 mg/mouse of hydrocortisone acetate by daily oral administration for 8 weeks to suppress host immune response.

3.5 Assessment of Infection and Treatment Efficacy

1. Mice are sacrificed at different time points post treatment.

2. Lungs and spleens from mice are removed rapidly by a sterile autopsy after sacrifice.

3. The organs are transferred into 2 mL tubes each containing 1 mL sterile distilled water and 2 mm diameter glass beads.

4. Lungs and spleens of mice are homogenized using a reciprocal shaker (Thermo Hybaid Ltd) for 40 s at 6.5 speed.

5. CFU counts and broth counts from each lung and spleen are performed using serial dilutions of the homogenates (see Subheading 3.2) and expressed as log CFU/organ for CFU counting or Log viable cells/organ for broth counting (see Subheading 3.6).

6. After 11 weeks of treatment, mouse organs are most likely to become CFU count free. Therefore, at the late stage of treatment, the entire organ homogenates (the total volume of each organ homogenate is approximately 1.5 mL including the organ and 1 mL of water) from 8 to 10 mice are aliquoted equally into three tubes.

7. Tube 1. CFU counting by addition of the homogenate to 2 mL of sterile distilled water following by plating out the entire organ homogenate suspension on six selective 7H11 agar plates (see **Note 12**).

8. Tube 2. Culturing in 5 mL of selective Kirchner liquid medium for 4 weeks with subsequent sub-culturing of the entire culture onto Löwenstein-Jensen slopes for a further 4 weeks (see **Note 13**).

9. Tube 3. Resuscitation of persistent bacteria (see Subheading 3.6).

10. Culture negative organs are defined as no colonies grown on 7H11 agar plates and no growth in selective Kirchner liquid medium following inoculation on Löwenstein-Jensen slopes.

11. After 8 weeks of hydrocortisone treatment, CFU counts from lungs and spleens are performed to determine disease relapse. The CFU counts will be performed by dilution of the organ homogenizes in a tenfold serial and plating 3×100 µL of all dilutions including the entire tissue homogenate on selective 7H11 agar plates (see **Note 14**).

3.6 Resuscitation of M. tuberculosis Persisters In Vitro and in Mice

Broth counting is performed as serial tenfold dilutions [7, 8].

1. Add 4.5 mL of the culture filtrates (see Subheading 3.1, **step 5**) in 7 mL Bijou tubes.

2. Add 0.5 mL of in vitro cultures or tissue homogenates in 4.5 mL culture filtrates in triplicate and mix well.

3. Then perform the culture in a tenfold serial dilution by adding 0.5 mL of the mixed cell suspension to 4.5 mL of culture filtrates (*see* **Note 15**).

4. Incubate the cultures at 37 °C without disturbance.

5. At 10-day intervals over a 2-month period of incubation, the broth cultures are examined for visible turbidity.

6. Plate *M. tuberculosis* in turbid tubes on 7H11 agar plates to confirm colonial morphology (*see* **Note 16**).

7. Estimate the MPN of viable bacilli from the patterns of positive and negative tubes [15].

8. Plate the cultures from turbid tubes on blood agar medium (Oxoid) and Sabouraud dextrose agar (Oxoid) to check the sterility of the culture free from bacterial and fungal contamination (*see* **Note 17**).

3.7 Statistical Analysis

1. Student's *t*-test is used to determine the difference of CFU counts and broth counts between the experimental groups.

2. The difference between the relapse rates is determined by Chi-square test and Fisher's exact test. *P*-values <0.05 are considered significant.

4 Notes

1. *M. tuberculosis* forms clumps when grown in 7H9 medium. The addition of Tween 80 is to prevent the formation of clumps. The reconstituted medium needs to be mixed well till the powder is completely dissolved before sterilization.

2. ADC and OADC are heat sensitive and need to be added in the medium after cooling to 55 °C.

3. In order to obtain consistent results, this slopes will be obtain from Becton Dickinson. On receipt, store slope tubes in the dark at 2–8 °C. Avoid freezing and overheating and minimize exposure to light. Media stored according to the labeled instruction may be inoculated up to the recommended incubation times before the expiration date.

4. In the in vitro model, OD value of 1–1.5 will be achieved at 15–20 days of incubation. It is extremely important that the bacteria are in the stage to actively produce RPF but there is still sufficient nutrient available in the medium for subsequence period of incubation.

5. For *M. tuberculosis* to grow in 7H9 medium, an inoculation of log-phase culture to the medium is essential. In the hypoxia model, log-phase growth is defined from 7 to 10 days of growth which represents about 2×10^7 CFU/mL.

6. *M. tuberculosis* tends to grow in clumps even in the detergent-containing 7H9, especially when a culture is incubated without disturbance for more than 30 days. In unagitated cultures, the clumps sink to the bottom of the container where they form solid pellicles. This makes the direct and accurate measurement of growth difficult and often the results are poorly reproducible. For growth rate determination and the uniform exposure of the bacilli to experimental treatment, it is very important that the bacilli are evenly dispersed, preferably as single cells with a minimum level of clumps. Larger clumps can be broken by vortexing the culture with 1 mm glass beads. The glass beads will be sterilized by autoclaving and added to the Universal tubes before CFU counting.

7. Fine clumps can be broken by sonication. As sonication produces aerosols, caps of tubes containing *M. tuberculosis* will be tightly screwed and sonicated in an ultrasonic water bath in the Biosafety Cabinet. The time for sonication will not exceed 5 min because the bacterial cells will be killed with longer sonication.

8. CFU counts are calculated as CFU/mL = colony counts × dilution factor × 10. For example, if 100 colonies are found on the plate for the dilution of 10^{-5}, viability will be calculated as $100 \times 100,000 \times 10 = 1 \times 10^8$ CFU/mL.

9. In order to keep *M. tuberculosis* virulence, the strain will be grown in a mouse for 2 weeks. The lung will be collected and CFU counts are performed. A single colony will be picked and streaked on 7H11 agar plates which will be incubated at 37 °C for 3 weeks. The bacteria grown on the agar plates will be collected and added in 10 mL of PBS. The clumps will be broken as Subheading 3.2, **steps 5** and **6**. The bacterial strain will be stored at −70 °C for subsequent animal inoculation. To determine the CFU counts prior to animal inoculation, viable counting is performed prior to freezing and once again after thawing.

10. Treatment must start not later than 3 weeks after infection as a high does bacterial infection (10^5 CFU/mouse) will cause mouse death if the mice are left untreated.

11. Rifampicin is administered 1 h before the other drugs to avoid drug to drug interactions [16].

12. At 16 weeks of treatment, organ CFU counts usually become negative. Therefore entire organs or one-third of organs need to be plate out to examine the number of bacterial cells in the organs on agar plates.

13. This step is to further confirm that there are no bacterial cells in the organs as CFU count negative bacilli may grow in liquid medium.

14. Disease relapse is measured by dividing the numbers of mice with CFU count positive organs with the total numbers of the mice. The difference in relapse rates based on CFU count detection of *M. tuberculosis* is analyzed using a Fisher's exact test between two experimental groups.

15. Normally for resuscitation of persisters in vitro cultures and in animal organs, 12 of 10-fold dilutions are needed to ensure the coverage of growth in culture filtrates.

16. In the turbid tubes, CFU counts will be performed by plating a serial of tenfold dilution of the cultures on 7H11 agar plates. The presence of *M. tuberculosis* will be confirmed by colony morphology from single colonies.

17. Turbid tubes may be due to contamination with other bacteria or fungi. 50 μL of the cultures from turbid tubes will be placed on blood agar or Sabouraud dextrose agar plate in triplicate followed by spreading the cultures on the plates. Blood agar plates will be incubated at 37 °C for 24 h and Sabouraud dextrose agar plates are incubated at 24 °C for 72–96 h.

Acknowledgments

This work was supported by the Innovative Medicines Initiative Joint Undertaking resources which are composed of financial contributions from the European Union's Seventh Framework Programme (FP7/2007-2013) and EFPIA companies' in-kind contribution (grant number 115337). This publication reflects only the authors' view. The European Commission is not liable for any use that may be made of the information herein.

References

1. MRC (1974) Controlled clinical trial of four short-course (6-month) regimens of chemotherapy for treatment of pulmonary tuberculosis. Lancet 2:1100–1106

2. RM MC Jr, McDermott W, Tompsett R (1956) The fate of *Mycobacterium tuberculosis* in mouse tissues as determined by the microbial enumeration technique. II. The conversion of tuberculous infection to the latent state by the administration of pyrazinamide and a companion drug. J Exp Med 104:763–802

3. RM MC Jr, Tompsett R (1956) Fate of *Mycobacterium tuberculosis* in mouse tissues as determined by the microbial enumeration technique. I. The persistence of drug-susceptible tubercle bacilli in the tissues despite prolonged antimicrobial therapy. J Exp Med 104:737–762

4. Lipworth S, Hammond RJ, Baron VO, Hu Y, Coates A, Gillespie SH (2016) Defining dormancy in mycobacterial disease. Tuberculosis (Edinb) 99:131–142

5. Mukamolova GV, Turapov O, Malkin J, Woltmann G, Barer MR (2010) Resuscitation-promoting factors reveal an occult population of tubercle bacilli in sputum. Am J Respir Crit Care Med 181:174–180

6. Turapov O, O'Connor BD, Sarybaeva AA, Williams C, Patel H, Kadyrov AS, Sarybaev AS, Woltmann G, Barer MR, Mukamolova GV (2016) Phenotypically adapted *Mycobacterium tuberculosis* populations from sputum are toler-

ant to first-line drugs. Antimicrob Agents Chemother 60:2476–2483

7. Hu Y, Liu A, Ortega-Muro F, Alameda-Martin L, Mitchison D, Coates A (2015) High-dose rifampicin kills persisters, shortens treatment duration, and reduces relapse rate in vitro and in vivo. Front Microbiol 6:641

8. Hu Y, Pertinez H, Ortega-Muro F, Alameda-Martin L, Liu Y, Schipani A, Davies G, Coates A (2016) Investigation of elimination rate, persistent subpopulation removal, and relapse rates of *Mycobacterium tuberculosis* by using combinations of first-line drugs in a modified Cornell mouse model. Antimicrob Agents Chemother 60:4778–4785

9. Gupta RK, Srivastava BS, Srivastava R (2010) Comparative expression analysis of rpf-like genes of *Mycobacterium tuberculosis* H37Rv under different physiological stress and growth conditions. Microbiology 156:2714–2722

10. Tufariello JM, Jacobs WR Jr, Chan J (2004) Individual *Mycobacterium tuberculosis* resuscitation-promoting factor homologues are dispensable for growth in vitro and in vivo. Infect Immun 72:515–526

11. Davies AP, Dhillon AP, Young M, Henderson B, McHugh TD, Gillespie SH (2008) Resuscitation-promoting factors are expressed in *Mycobacterium tuberculosis*-infected human tissue. Tuberculosis (Edinb) 88:462–468

12. Mukamolova GV, Murzin AG, Salina EG, Demina GR, Kell DB, Kaprelyants AS, Young M (2006) Muralytic activity of *Micrococcus luteus* Rpf and its relationship to physiological activity in promoting bacterial growth and resuscitation. Mol Microbiol 59:84–98

13. McCune RM, Feldmann FM, McDermott W (1966) Microbial persistence. II. Characteristics of the sterile state of tubercle bacilli. J Exp Med 123:469–486

14. McCune RM, Feldmann FM, Lambert HP, McDermott W (1966) Microbial persistence. I. The capacity of tubercle bacilli to survive sterilization in mouse tissues. J Exp Med 123:445–468

15. Sutton S (2010) The most probable number method and its uses in enumeration, qualification and validation. J Valid Tech 16:4

16. Grosset J, Truffot-Pernot C, Lacroix C, Ji B (1992) Antagonism between isoniazid and the combination pyrazinamide-rifampin against tuberculosis infection in mice. Antimicrob Agents Chemother 36:548–551

INDEX

Stephen H. Gillespie (ed.), *Antibiotic Resistance Protocols*, Methods in Molecular Biology, vol. 1736,
https://doi.org/10.1007/978-1-4939-7638-6, © Springer Science+Business Media, LLC 2018

Printed in the United States
By Bookmasters